AF342590

REPUBLIC OF SOUTH AFRICA
REPUBLIEK VAN SUID-AFRIKA

DEPARTMENT OF AGRICULTURE AND WATER SUPPLY
DEPARTEMENT VAN LANDBOU EN WATERVOORSIENING

FLORA OF SOUTHERN AFRICA

VOLUME 4

PART 2

ISBN 0 621 08271 6

G.P.-S

FLORA OF SOUTHERN AFRICA

which deals with the territories of

SOUTH AFRICA, CISKEI, TRANSKEI, LESOTHO, SWAZILAND, BOPHUTHA-TSWANA, SOUTH WEST AFRICA/NAMIBIA, BOTSWANA AND VENDA

VOLUME 4

PART 2

XYRIDACEAE–JUNCACEAE

by

A. A. Obermeyer, John Lewis & Robert B. Faden

Edited by

O. A. Leistner

Editorial Committee: B. de Winter, D. J. B. Killick and O. A. Leistner

Botanical Research Institute,
Department of Agriculture
and Water Supply
1985

CONTENTS

NEW COMBINATION PUBLISHED IN PART 2*

Eriocaulon dregei *Hochst.* var. **sonderianum** *(Koern.) Oberm.*, comb., nov., p.2: 18.

*Date of publication: April, 1985

INTRODUCTION

Keys to families are provided in R. A. Dyer's Genera of Southern African Flowering Plants, Vol. 1 (1975) which is arranged on the lines of the Engler system. The genera are numbered, as far as possible, according to the list published by De Dalla Torre and Harms in their Genera Siphonogamarum (1900–1907) in order to facilitate reference, though genera in the Flora are not necessarily arranged in this sequence.

The following condensed abbreviations for literature references are used:

C.F.A.	Conspectus Florae Angolensis
R. A. Dyer, Gen.	The Genera of Southern African Flowering Plants by R. A. Dyer, Vol. 1 (1975) and Vol. 2 (1976)
F.C.	Flora Capensis
F.C.B.	Flore du Congo et du Rwanda-Burundi
F.S.W.A.	Prodromus einer Flora von Südwestafrika
F.T.A.	Flora of Tropical Africa
F.T.E.A.	Flora of Tropical East Africa
F.W.T.A.	Flora of West Tropical Africa
F.Z.	Flora Zambesiaca

This fascicle was compiled in accordance with a Guide to Contributors to the Flora of Southern Africa (Ross, Leistner & De Winter, 1977), which is available from the Librarian, Botanical Research Institute, Private Bag X101, Pretoria 0001.

The family Mayacaceae, which was included in R. A. Dyer, Gen. 1: 905 (1975), and the genus *Triceratella* Brenan (in Kirkia 1: 14; 1961) of the Commelinaceae given on p. 910 of the same work, have been excluded here, as no confirmed records of representatives were seen from our region.

Volume 4 of the Flora, of which the present publication is a component, will appear in two parts (see p. viii). The number of the part, which in the present publication is '2', precedes the page number on all pages marked with arabic numerals. This was done with a view to compiling a combined index to the entire volume.

PLAN OF FLORA OF SOUTHERN AFRICA

Cryptogam volumes will in future not be numbered but will be known by the name of the group they cover. The number assigned to the volume on Charophyta therefore becomes redundant.

Alien families are marked with an asterisk.

Published volumes and parts are shown in italics.

Please note that local prices as given below do not include GST. Prices given for other countries include postage.

INTRODUCTORY VOLUMES

The genera of Southern African flowering plants
    Vol. 1: *Dicotyledons* (Published 1975). Price: R11,23. Other Countries: R14,00
    Vol. 2: *Monocotyledons* (Published 1976). Price: R8,21. Other countries: R10,00

Botanical exploration in Southern Africa (Published 1981). Price: R40,00 (Obtainable from booksellers)

CRYPTOGAM VOLUMES

Charophyta (Published as Vol. 9 in 1978). Price: R4,25. Other countries: R5,30

Bryophyta:
    Part 1: Mosses: Fascicle 1: *Sphagnaceae – Grimmiaceae* (Published 1981). Price: R24,34. Other countries: R30,40
    Fascicle 2: Gigaspermaceae – Bartramiaceae
    Fascicle 3: Erpodiaceae – Hookeriaceae
    Fascicle 4: Fabroniaceae – Polytrichaceae

Pteridophyta

FLOWERING PLANTS VOLUMES

Vol. 1: *Stangeriaceae, Zamiaceae, Podocarpaceae, Pinaceae*, Cupressaceae, Welwitschiaceae, Typhaceae, Zosteraceae, Potamogetonaceae, Ruppiaceae, Zannichelliaceae, Najadaceae, Aponogetonaceae, Juncaginaceae, Alismataceae, Hydrocharitaceae* (Published 1966). Price: R1,98. Other countries: R2,60

Vol. 2: Poaceae

Vol. 3: Cyperaceae, Arecaceae, Araceae, Lemnaceae, Flagellariaceae

Vol. 4: Part 1: Restionaceae
    Part 2: *Xyridaceae, Eriocaulaceae, Commelinaceae, Pontederiaceae, Juncaceae* (Published 1985). Price: R7,50. Other countries: R9,40

Vol. 5: Liliaceae, Agavaceae

Vol. 6: Haemodoraceae, Amaryllidaceae, Hypoxidaceae, Tecophilaeaceae, Velloziaceae, Dioscoreaceae

Vol. 7: Iridaceae: Part 1: Nivenioideae, Iridoideae
    Part 2: Ixioideae: Fascicle 1
    Fascicle 2: *Syringodea, Romulea* (Published 1983)
    Price: R3,96. Other countries: R5,00

Vol. 8: Musaceae, Strelitziaceae, Zingiberaceae, Cannaceae*, Burmanniaceae, Orchidaceae

Vol. 9: Casuarinaceae*, Piperaceae, Salicaceae, Myricaceae, Fagaceae*, Ulmaceae, Moraceae, Cannabaceae*, Urticaceae, Proteaceae

Vol. 10: Part 1: *Loranthaceae, Viscaceae* (Published 1979). Price: R4,34. Other countries: R5,40
Santalaceae, Grubbiaceae, Opiliaceae, Olacaceae, Balanophoraceae, Aristolochiaceae, Rafflesiaceae, Hydnoraceae, Polygonaceae, Chenopodiaceae, Amaranthaceae, Nyctaginaceae

Vol. 11: Phytolaccaceae, Aizoaceae, Mesembryanthemaceae

Vol. 12: Portulacaceae, Basellaceae, Caryophyllaceae, Illecebraceae, Cabombaceae, Nymphaeaceae, Ceratophyllaceae, Ranunculaceae, Menispermaceae, Annonaceae, Trimeniaceae, Lauraceae, Hernandiaceae, Papaveraceae, Fumariaceae

Vol. 13: *Brassicaceae, Capparaceae, Resedaceae, Moringaceae, Droseraceae, Roridulaceae, Podostemaceae, Hydrostachyaceae* (Published 1970). Price: R10,00. Other countries: R12,00

Vol. 14: Crassulaceae (in press)

Vol. 15: Vahliaceae, Montiniaceae, Escalloniaceae, Pittosporaceae, Cunoniaceae, Myrothamnaceae, Bruniaceae, Hamamelidaceae, Rosaceae, Connaraceae

Vol. 16: Fabaceae: Part 1: *Mimosoideae* (Published 1975). Price: R13,59. Other countries: R16,75
Part 2: *Caesalpinioideae* (Published 1977). Price: R16,04. Other countries: R20,00
Papilionoideae
Vol. 17: Geraniaceae, Oxalidaceae
Vol. 18: Linaceae, Erythroxylaceae, Zygophyllaceae, Balanitaceae, Rutaceae, Simaroubaceae, Burseraceae, Ptaeroxylaceae, Meliaceae, Aitoniaceae, Malpighiaceae
Vol. 19: Polygalaceae, Dichapetalaceae, Euphorbiaceae, Callitrichaceae, Buxaceae, Anacardiaceae, Aquifoliaceae
Vol. 20: Celastraceae, Icacinaceae, Sapindaceae, Melianthaceae, Greyiaceae, Balsaminaceae, Rhamnaceae, Vitaceae
Vol. 21: Part 1: *Tiliaceae* (Published 1984). Price: R4,30. Other countries: R5,00
Malvaceae, Bombacaceae, Sterculiaceae
Vol. 22: *Ochnaceae, Clusiaceae, Elatinaceae, Frankeniaceae, Tamaricaceae, Canellaceae, Violaceae, Flacourtiaceae, Turneraceae, Passifloraceae, Achariaceae, Loasaceae, Begoniaceae, Cactaceae* (Published 1976). Price: R8,68. Other countries: R10,75
Vol. 23: Geissolomaceae, Penaeaceae, Oliniaceae, Thymelaeaceae, Lythraceae, Lecythidaceae
Vol. 24: Rhizophoraceae, Combretaceae, Myrtaceae, Melastomataceae, Onagraceae, Trapaceae, Haloragaceae, Gunneraceae, Araliaceae, Apiaceae, Cornaceae
Vol. 25: Ericaceae
Vol. 26: *Myrsinaceae, Primulaceae, Plumbaginaceae, Sapotaceae, Ebenaceae, Oleaceae, Salvadoraceae, Loganiaceae, Gentianaceae, Apocynaceae* (Published 1963). Price: R4,53. Other countries: R5,75
Vol. 27: Part 1: Periplocaceae, Asclepiadaceae (Microloma – Xysmalobium)
Part 2: Asclepiadaceae (Schizoglossum – Woodia)
Part 3: Asclepiadaceae (Asclepias – Anisotoma)
Part 4: *Asclepiadaceae (Brachystelma – Riocreuxia)* (Published 1980). Price: R4,43. Other countries: R6,00
Asclepiadaceae (remaining genera)
Vol. 28: Part 1: Cuscutaceae, Convolvulaceae
Part 2: Hydrophyllaceae, Boraginaceae
Part 3: Stilbaceae, Verbenaceae
Part 4: Lamiaceae (in press)
Part 5: Solanaceae, Retziaceae
Vol. 29: Scrophulariaceae
Vol. 30: Bignoniaceae, Pedaliaceae, Martyniaceae, Orobanchaceae, Gesneriaceae, Lentibulariaceae, Acanthaceae, Myoporaceae
Vol. 31: Plantaginaceae, Rubiaceae, Valerianaceae, Dipsacaceae, Cucurbitaceae
Vol. 32: Campanulaceae, Sphenocleaceae, Lobeliaceae, Goodeniaceae
Vol. 33: Asteraceae: Part 1: Lactuceae, Mutisieae, 'Tarchonantheae'
Part 2: Vernonieae, Cardueae
Part 3: Arctotideae
Part 4: Anthemideae
Part 5: Astereae
Part 6: Calenduleae
Part 7: Inuleae: Fascicle 1: Inulinae
Fascicle 2: *Gnaphaliinae (First part)* (Published 1983). Price: R12,93. Other countries: R16,20
Part 8: Heliantheae, Eupatorieae
Part 9: Senecioneae

XYRIDACEAE

by John Lewis* and A. A. Obermeyer

Herbs, annual or perennial, usually scapigerous from tufted bases and sometimes with swollen or rhizomatous rootstocks, usually gregarious. *Leaves* linear or filiform, usually sheathing at base. *Inflorescence* usually spicate, formed of imbricated bracts in an ovoid head terminating a scape, very rarely a few-flowered cyme; outer 2 to few pairs of bracts often sterile and sometimes involucrate. *Flowers* bisexual, regular or partially zygomorphic. *Sepals* 3, or rarely 2, lower 2 placed laterally and opposite. *Petals* 3, free or united, regular. *Stamens* 3, opposite and adnate to petals, often with 3 alternate staminodes; anthers extrorse. *Ovary* superior, usually unilocular; placentation parietal, basal or free-central; ovules orthotropous, numerous; seeds usually apiculate and longitudinally striate, albuminous.

A family comprising four genera: two smaller ones confined to the Guayana highland of South America, one (*Abolboda* H.B.K.) with about 15 spp. in north-eastern South America and the large genus *Xyris* itself with about 200 species, widespread in warmer regions of Africa, Asia, Australia and particularly in the Americas, but absent from Europe. In Southern Africa 7 species are recorded in the northern areas, one of which is widespread, extending to the southern Cape; they inhabit marshes, seepage areas, riverbanks or moist rock crevices.

826   XYRIS

Xyris *L.*, Sp. Pl. 42 (1753), Gen. Pl. edn 5: 25 (1754); Benth. & Hook. f., Gen. Pl. 3: 842 (1883); N. E. Br. in F.C. 7: 3 (1897), and in F.T.A. 8: 7 (1901); Malme in Bot. Jb. 48: 287 (1912), and in Natürl. PflFam. edn 2, 15a: 35 (1930); Roessl. in F.S.W.A. 158 (1967); R. A. Dyer, Gen. 2: 906 (1976); Lewis in Fl. Cameroun 22: 35 (1981). Type species: *Xyris indica* L.

Herbs, annual or perennial and caespitose, rush-like, scapigerous. *Leaves* radical, spirally arranged, tufted, linear or filiform, isolateral, more or less sheathing at base. *Scape* naked, usually much exserted above leaves, sheathed below, sheath open, terminated by a short caudicle. *Inflorescence* spicate; bracts usually tightly imbricate, sometimes marked with a median contrasting area on dorsal surface, outer 1 or 2 pairs sterile. *Sepals* 3; lateral sepals opposite and more or less conduplicate, chartaceous, usually keeled, keel varied in extent and ornamentation; median sepal forming a caducous membranous hood. *Petals* free below, yellow, rarely white and even in one case blue or violet. *Staminodes* usually present, often bipartite. *Ovary* with a variable placentation; style usually 3-branched. *Capsule* loculicidal, splitting between parietal placentas (in Southern African species); seeds numerous, albuminous, minute, ellipsoid, pointed on both sides with longitudinal, raised, often moniliform ribs, connected by thin cross bars and with secondary thinner ribs in between.

Seven species recorded from Southern Africa, none endemic; several of these widespread in the warmer parts of Africa and elsewhere. A genus of 3 sections of which only the section *Xyris* occurs in Africa, characterized by the possession of a unilocular capsule with parietal placentation.

The broad species concept accepted here may be clarified by reference to the detailed synonymies for five of the species dealt with by Lewis (loc. cit.).

Name derived from the Greek in reference to the leaves possessing razor-sharp margins.

1 Leaves distichous, 2–8, from a small base; annuals, or, if perennial, with an inconspicuous rhizome; keels of lateral sepals smooth, not ciliate:
 2 Scape terete, green above, lower half golden brown and shiny; spikes c. 6-flowered, bracts shiny brown; (included are paedogenic small plants, 2-leaved, 1–3-flowered, the bracts membranous, often vinaceous); widespread .... 1. *X. capensis*
 2 Scape ancipitous, uniformly green; spikes many-flowered; bracts stramineous; coarse annuals up to 0,4 m tall; a mostly coastal tropical species extending south to Natal and north-eastern Cape .... 2. *X. anceps*

---

* Formerly of Department of Botany, British Museum (Natural History); now, International Registrar for Conifers (R.H.S. appointed), 83 High Street Hampton Hill, Middx TW12 1NH, U.K.

1 Leaves clustered on compact rhizomes; cataphylls, leaf-bases and peduncular sheaths sclerotic, shiny dark brown; keels of lateral sepals ciliate:
  3 Leaves terete; outer lower bracts of conical spike often somewhat longer, patent and paler than inner; roots hard, thick .... 3. *X. natalensis*
  3 Leaves linear; bracts of spike more or less uniform; roots usually thin to fairly thin:
    4 Spikes ovoid, bracts dark brown, in a fairly uniform spiral:
      5 Tall plants up to 0,5 m; spikes many-flowered .... 4. *X. congensis*
      5 Smaller plants up to c. 0,3 m; spikes few-flowered; the base of the plant often sub-bulbous, fibrous, twisted, mid-brown; roots thin .... 5. *X. nivea*
    4 Spikes obovoid, obtuse to broad apically; lower bracts larger than upper:
      6 Bracts with scarious splitting margins; spikes c. 6-flowered .... 6. *X. gerrardii*
      6 Bracts with entire margins, not scarious, fertile, often with a small greyish median patch below apex; spikes c. 20-flowered .... 7. *X. rehmannii*

1. **Xyris capensis** *Thunb.*, Prodr. 12 (1794), Fl. Cap. edn 2: 81 (1823); N. E. Br. in F.C. 7: 6 (1897), and in F.T.A. 8: 13 (1901); Adamson in Adamson & Salter, Fl. Cape Penins. 159 (1950); Roessl. in F.S.W.A. 158: 1 (1967); Ellis, Manders & Oberm. in Bothalia 12: 637 (1979); Lewis in Fl. Cameroun 22: 37 (1981). Type: Cape, near Verkeerde valley, *Thunberg* 1267 (UPS, holo., PRE, photo.!).

X. rubella Malme in Bot. Jb. 48: 303 (1912); Roessl. in F.S.W.A. 158: 2 (1967); Hepper in Kew Bull. 21: 424 (1968). Syntypes: South West Africa/Namibia, Okahandja, *Dinter* 944 (B!; S!; SAM!); Tanzania, between Orero and Kilwa Karingi, *Braun* in herb. Amani 1326 (EA, lecto.!; S!). Considered a paedogenic variant (see notes below).

Tufted perennials up to c. 0,3 m tall, but also paedogenic, producing flowers when barely past seedling stage. *Rhizome* small, roots bilaterally arranged, many. *Leaves* opposite, erect, linear, laterally flattened, c. 50–150 × 4 mm, broadening slightly from base and tapering to an acute apex, terminating in a small obtuse callosity; cataphylls 0. *Scapes* about twice as long as leaves, usually terete, green and minutely papillate at first but, with subsequent lengthening, basal portion loses outer epidermis becoming golden-brown, smooth and shiny. *Spikes* ellipsoid to rounded, c. 10 mm in diam. (1)–3–5-flowered; bracts numerous, rather loosely imbricate at maturity, margins recurved or divergent, olive brown, subtranslucent; outer 2 pairs broadly elliptic, hooded or flat, not keeled, inner more or less straight and conduplicate, keeled, apex sharply acute. *Lateral sepals* slightly curved, hyaline, keeled, apex acute, keel winged in lower half, margin entire or rarely with a very few broad teeth. *Staminodes* scarcely exserted from corolla-tube. *Capsule* oblong-ovoid, 3–4 mm long, obtuse; seeds ellipsoid, 0,5 mm long. Fig. 1: 2.

A widespread and very common pioneer species in Southern Africa as far south as the Cape Peninsula; also in Madagascar and tropical Africa; further in South America and India, Sri Lanka, China and Malaysia; in marshes, on river banks and grassy seepage areas. Map 1.

Vouchers: *Acocks* 20416; *Burtt & Hilliard* 8715; *Dieterlen* 602; *Galpin* 9096; *Rehmann* 7364; *Rudatis* 1231; *Schlechter* 289.

Plants in sites which are only seasonally wet, remain small and ephemeral. Stunted plants no higher than 40–60 mm, with only two filiform leaves, will often produce inflorescences with only one to two flowers.

Plants in permanently wet localities will develop to maturity and become perennial, bearing rhizomes and producing new sets of distichous leaves bilaterally from the centre of the plant.

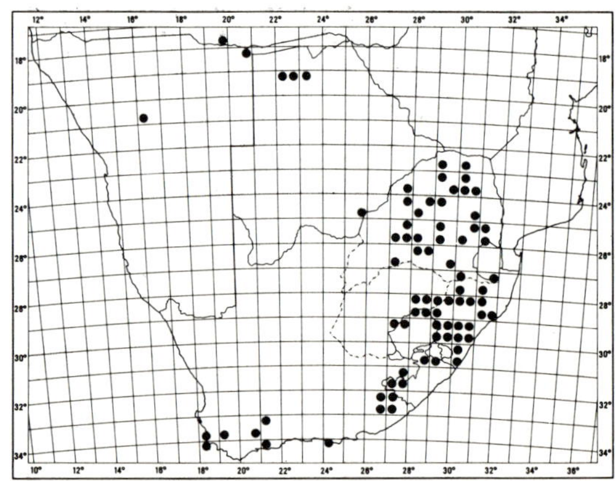

MAP 1.— **Xyris capensis**

2. **Xyris anceps** *Lam.*, Tab. Encycl. Meth. Bot. 1: 132 (1791); N. E. Br. in F.C. 7: 6 (1897), and in F.T.A. 8: 12 (1901); Lewis in Fl. Cameroun 22: 42 (1981). Type: Madagascar, *Commerson* s.n. (P, holo.; BM, iso.!).

Soft herbs up to c. 0,7 m tall. *Leaves* distichously arranged, each shoot consisting of a scape enveloped basally by 1–2 leaves and a sheath, linear, up to 240 × 5 mm, apex acute.

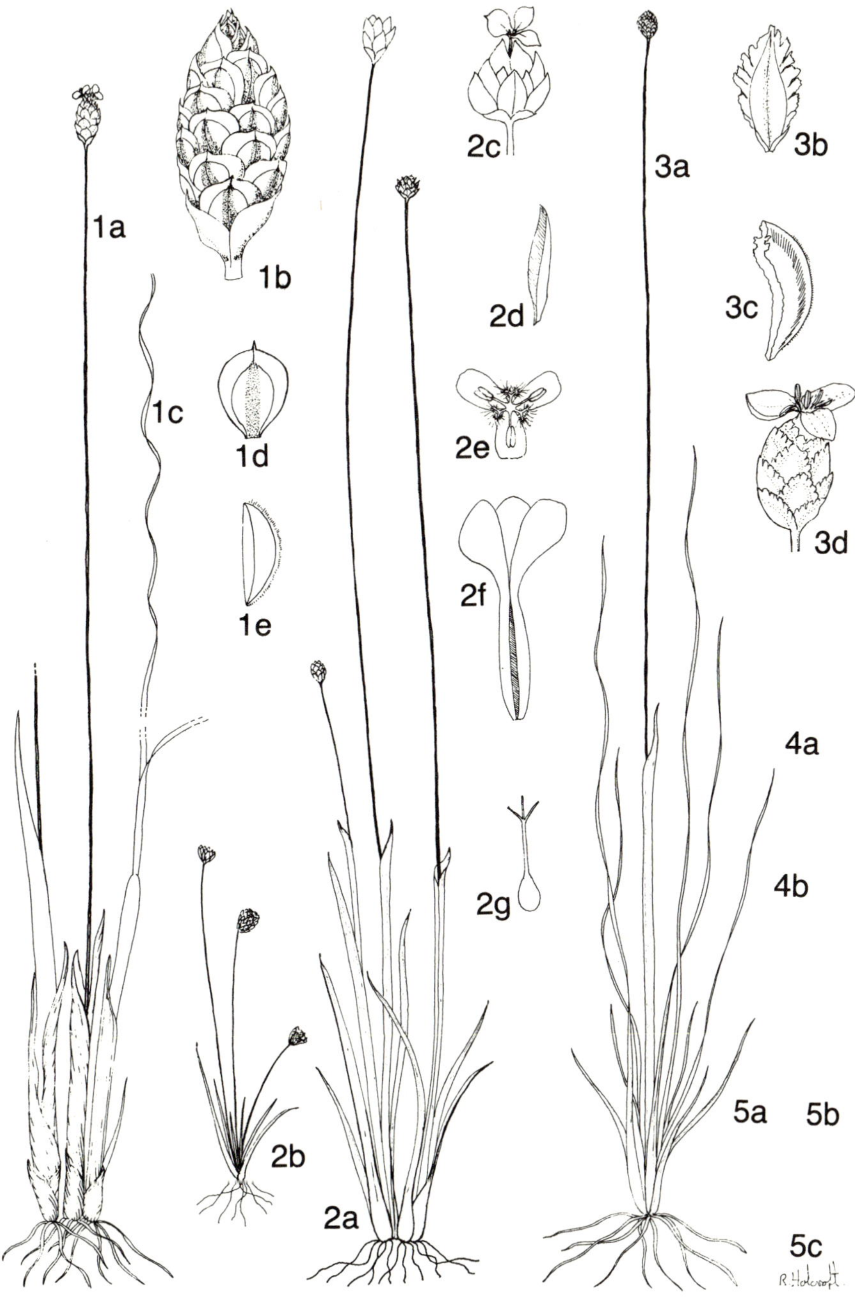

FIG. 1.—1, **Xyris congensis**: 1a, habit, × 0,5; 1b, capitulum, × 2; 1c, twisted leaf blade, × 0,5; 1d, bract, × 3; 1e, lateral sepal with ciliate keel, × 3 (*Mauve* 5031). 2, **X. capensis**: 2a, habit, 0,5; 2b, paedogenic stage ("*X. rubella*"), × 0,5; 2c, capitulum, × 1,5; 2d, lateral sepal, × 3; 2e, petals, stamens and staminodes, × 2,5; 2f, fused claws of petals splitting when ovary develops, × 3; 2g, ovary, style and stigmas, × 2 (*Buitendag* 409, 724). 3, **X. gerrardii**: 3a, habit, × 0,5; 3 b, floral bract showing lacerated membranous margin, × 3; 3c, lateral sepal with ciliate keel, × 3; 3d, capitulum, × 1,5 (*Mauve* 5032).

Scapes 2–3 times as long as leaves, strongly two-edged, especially so above; sheath resembling leaf but shorter. *Spikes* broadly ellipsoid to spherical, up to c. 14 mm tall, many-flowered. *Bracts* numerous and rather regularly spirally arranged, broadly ovate, acute, convex, shiny, yellow to olive green, with a grey diamond-shaped depression below acute apex, often splitting with age. *Lateral sepals* short and curved, folded, with a glabrous entire keel. *Perianth* typical. *Capsule* flattened, oblong, obtuse; seeds ellipsoid, 0,3–0,5 mm, typical.

Recorded from the south-eastern Transvaal (only once!), Natal and Transkei, as far south as Pondoland; also in subtropical to tropical Africa, Mascarene Islands and in Brazil; commonly inhabiting coastal swamps. Map 2.

Vouchers: *Acocks* 13337; *Gordon-Gray* 6167; *Rehmann* 8560; *Tinley* 3061; *Ward* 7717; *Wood* 11989.

Not as variable a species as its synonymy from outside our area would suggest; readily recognized by its pale green colour and sharply 2-angled scape.

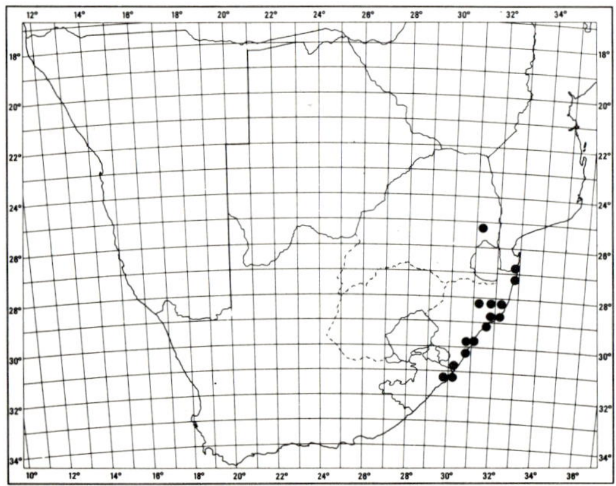

MAP 2.— **Xyris anceps**

3. **Xyris natalensis** *Nilsson* in Öfvers. K. Vet. Akad. Förhandl. 1891: 157 (1891), and in K. svenska VetensAkad. Handl., ny följd 24, 14: 28 (1892); Durand & Schinz, Consp. Fl. Afr. 5: 421 (1895); N. E. Br. in F.C. 7: 4 (1897). Syntypes: Natal, Natal Bay, *Krauss* (141) (M, putative lecto.!; BM!; K!; Z!); Natal without precise locality, *Sanderson* 455 (wrongly cited by Nilsson as 456) (S; K!); Cape, without precise locality, *Wahlberg* s.n. (S).

X. foliolata Nilsson in K. svenska VetensAkad. Handl., ny följd 24, 14: 65 (1892); Durand & Schinz, Consp. Fl. Afr. 5: 420 (1895); N. E. Br. in F.T.A. 8: 10

(1901); Malme in Ark. Bot. 24A: 2 (1932). Type: Angola, Malange, *Mechow* s.n. (Z, holo.!).

Tufted perennials up to 0,85 m tall. *Rhizome* subhorizontal, hard, compact, covered by persistent, shiny, brown leaf-bases and cataphylls; roots thick, hard. *Leaves* clustered, terete, up to 500 mm long, c. 1 mm thick, wiry, apex acute, widened abruptly below to form an open, auriculate sheath. *Spikes* on exserted, terete, wiry peduncles enclosed basally by an open, cylindrical sheath; heads often cone-like, ovoid to oblong-ovoid, c. 10–28 mm long, mid-brown, shiny; lowest sterile bracts very narrowly ovate, smooth, frequently aristate, fertile ones with a minutely punctate median area, margin entire or mildly erose. *Lateral sepals* folded, falcate, apiculate; keel prominent, densely ciliate especially in upper half. *Capsule* narrowly cylindrical, 3 mm long, thin-walled. *Seeds* typical.

Natal to Transkei, confined to the coastal areas; known also from Angola, Burundi, Zambia and Zimbabwe. Common in swamps, on river banks and along lakes. Flowering in summer. Map 3.

Vouchers: *Hilliard* 3182; *Gordon-Gray* 6170; *Strey* 4488, 4905; *Venter* 929; *Ward* 2857; *Wood* 9939.

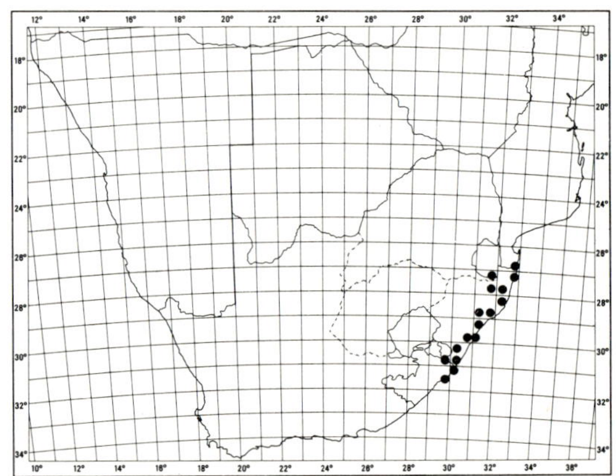

MAP 3.— **Xyris natalensis**

4. **X. congensis** *Buettn.* in Verh. bot. Ver. Prov. Brandenb. 31: 71 (1889); N. E. Br. in F.T.A. 8: 23 (1901); Lewis in Fl. Cameroun 22: 46 (1981). Type: Zaïre, Congo River between Lukolela and Equatorville, *Büttner* 583 (B, holo.!, PRE, photo.!).

X. umbilonis Nilsson in K. svenska VetensAkad. Handl., ny följd 24, 14: 30 (1892); N. E. Br. in F.C. 7: 4 (1897). Type: Natal, Mbilo waterfall, *Rehmann* 8139 (Herb. Schinz in Z, holo.!; Z, iso.!; BM, iso.!).

X. batokana N. E. Br. in F.T.A. 8: 22 (1901). Type: Zambia, Batoka highlands, *Kirk* s.n. (K, holo.!).

Perennial, strongly tufted; rhizome hard, compact, covered by persistent shiny reddish brown contorted leaf-bases and cataphylls up to 60 mm long; roots fairly thin, soft. *Leaves* clustered, stiff, linear, up to 500 × 4 mm (less in our area), acuminate above, sometimes mildly twisted. *Scapes* up to 0,6 m long, wiry, terete or flattened when dry. *Spikes* ellipsoid, up to 15 × 7 mm, bracts fairly uniform and somewhat gaping at anthesis, shining chestnut-brown, convex, mucronate, margin smooth, entire except for an occasional median split. *Lateral sepals* folded, curved with thick dark keels bearing a narrow wing more or less densely ciliate, especially above. *Calyptriform sepal* brownish yellow. *Corolla* yellow, fading to white, lobes of petals denticulate. *Style* with 3 capitate stigmas. *Capsule* narrowly ovoid, 2–3 mm long; seeds ovoid. Fig. 1: 1.

As conceived here, a species widely spread in Southern Africa from Transvaal through Swaziland to Natal, and further north in tropical Africa from west Africa eastwards to Tanzania and Madagascar. A plant of damp areas, often beside streams, forming large flourishing tufts; often in montane areas, sometimes dominant in swampy conditions. Map 4.

Vouchers: *Bos* 1087; *Galpin* 1223, 9097; *Mauve & Venter* 5201; *Pegler* 1018; *Rogers* 24991; *Rudatis* 397.

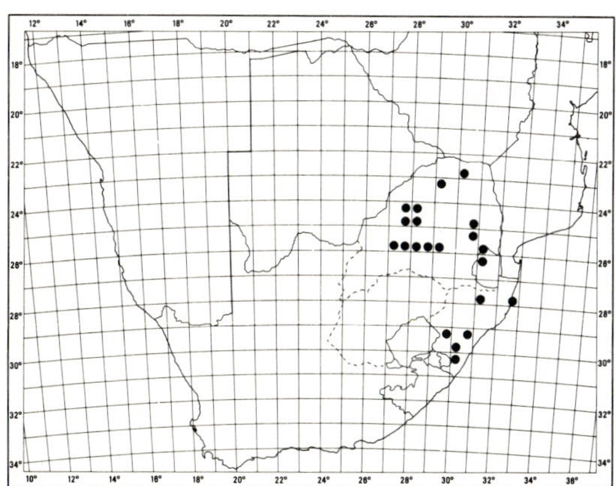

MAP 4.— **Xyris congensis**

5. **Xyris nivea** *Welw. ex Rendle*, Cat. Afr. Pl. Welw. 2, 1: 69 (1899); J. Lewis in Fl. Cameroun 22: 40 (1981). Type: Angola, Huilla, between Lopollo and Monino, *Welwitsch* 2468 (BM, holo.!, PRE, photo.!).

Xyris obscura N.E. Br. in F.T.A. 8, 1: 16 (1901, June); Malme in Bot. Jb. 48: 288, 301 (1912), *in obs. sub X. brunnea*; Hepper in Kew Bull. 21: 424 (1968); Type: Zim-babwe, near Salisbury [Harare], Six-mile Spruit, *Mrs Cecil* 152 A(K, holo.!).

Xyris brunnea Nilsson in Bot. Jb. 30: 271 (1901, July); Malme in Bot. Jb. 48: 300 (1912), and in Bot. Notiser 1932: 12 (1932). Type: Tanzania, Livingstone Mts, western Ubena, *Goetze* 822 (B, holo.!; BM, iso.!; EA, iso.!).

Xyris aberdarica Malme in Notizbl. bot. Gart. Mus. Berl. 8: 663 (1924). Type: Kenya, Mt Aberdare, near Kinangop Forest Station, *R. E. & T. C. E. Fries* 2905 (UPS, holo.!; K, iso.!; B, iso.!).

Xyris sp. near *obscura* N.E. Br., sensu Hepper in F.W.T.A. edn 2, 3: 54 (1968).

Perennial, with base inflated and sometimes sub-bulbous. *Leaves* somewhat stiff, very narrow and even appearing terete (especially from burnt culms) but characteristically flat and parallel-sided at maturity, up to 180 mm long, c. 1 mm wide, frequently tortuous, minutely cross-striate, especially at midlength. *Scapes* (1–)3 times as long as leaves, terete and multistriate, frequently irregularly compressed, green or obscurely pale brown at base (within sheath); sheaths often short and obscure, especially on high mountain plants, with a lanceolate caudicle about 10 mm long, or sheaths sometimes ⅓ to ½ times as long as taller scapes, and broader, rarely to as much as 4–5 mm wide. *Capitula* broadly ellipsoid or sub-spherical, 6–8 mm long, few-flowered. *Bracts* numerous, rather loosely imbricate when mature, tough, brittle, more or less concave and medium to dark brown; outer elliptic, more or less flat, frequently keeled and with or without a hard mucro or seta; inner somewhat broader, more often keeled and usually with lighter coloured, frequently recurved margins; apices obtuse. *Lateral sepals* slightly curved, even sometimes mildly recurved above, hyaline, keeled; apices sometimes blunt, usually acute, hard, black, with or without short terminal setae; keels moderately wide, sometimes more or less uniformly so but usually diminishing below and above, regularly and somewhat distantly shortly pubescent or puberulent, especially in mid-part, diminishing towards apices. *Flowers* yellow, very rarely white (*Welwitsch* 2468). Capsule and seeds not seen.

Recorded from the central and north-eastern Transvaal; rare. Widely distributed in central Africa from Angola and Mozambique northwards to Kenya; a species of wet areas in upland rocky steppes; also more rarely on lake-side marshes. Map 5.

Vouchers: *Junod* 2758; *Lavranos* 9366; *Prosser* 1797; *Rogers* 14930; *Smuts & Gillett* 2270; *Stirton* 224; *Werdermann & Oberdieck* 2075.

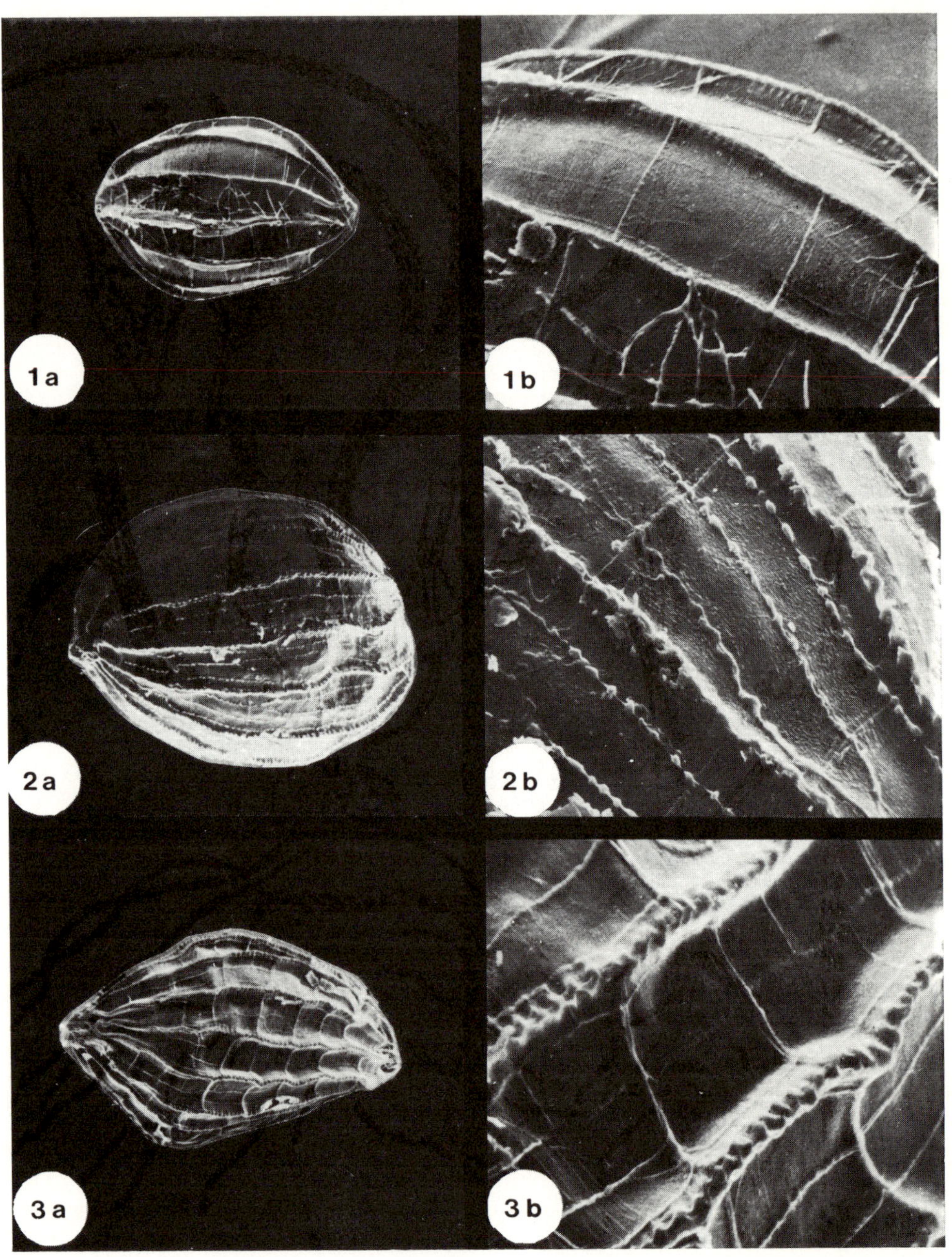

PLATE 1.—Seeds of **Xyris** species: 1, **X. anceps:** 1a, 120; 1b × 420 (*De Winter & Vahrmeijer* 8560). 2, **X. gerrardii:** 2a, × 120; 2b, × 600 (*Downing* 74). 3, **X. rehmannii:** 3a, × 120; 3b, × 600 (*Van Vuuren* 1324).

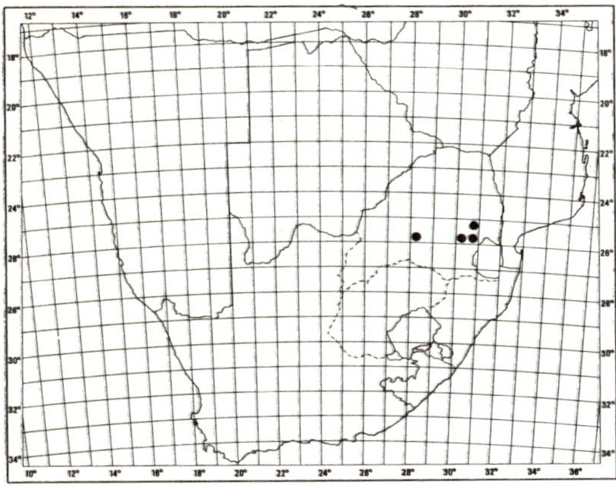

MAP 5.— **Xyris nivea**

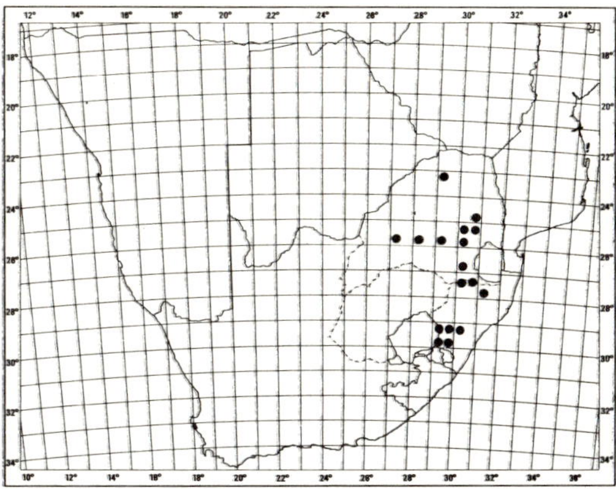

MAP 6.— **Xyris gerrardii**

6. **X. gerrardii** *N.E. Br.* in F.C. 7: 5 (1897); Malme in Bot. Jb. 48: 299 (1912); Malme apud Norlind & Weimarck in Bot. Notiser 1932: 12 (1932). Type: Natal, Zululand, *Gerrard* 1526 (K, holo.!; NH, iso.!; BM, iso.!).

Tufted perennials up to c. 0,3 m tall; rhizome hard, compact, densely covered with dark, hard shiny sheathing leaf-bases and cataphylls up to c. 50 mm long; roots thin. *Leaves* grass-like, numerous, linear, 0,35 m long and 1 mm wide, soft, thin. *Spikes* on long thin peduncles up to c. 0,55 m long; ellipsoid to obovoid, 5–8-flowered, up to 7,5 mm long and 5 mm in diam. *Bracts* strongly carinate above, acute, 5–7 mm long, dark shiny brown with a broad scarious margin, lacerate with age. *Flowers*: several often flowering simultaneously on one spike. *Lateral sepals* obovate-attenuate, keeled, keel not winged, fringed with a short distinct puberulence, especially medially, rarely glabrous. *Calyptriform sepal* deep orange-red. *Petals* yellow, lobes rotundate, coarsely dentate. *Stamens* and staminodes typical. *Stigmas* dilated, fimbriate. *Capsule* oblong ovoid, c. 4 mm long; seeds typical. Fig. 1: 3. Plate 1: 2.

Recorded from Transvaal and Natal; also in Zimbabwe; mostly in colder montane swamps. Map 6.

Vouchers: *Codd & Dyer* 9076; *Hilliard & Burtt* 7903, 9350, 9413, 8738; *Gilfillan* 401; *Mauve* 5032, 5263; *Rand* 89, 1045; *Schlechter* 3815.

7. **Xyris rehmannii** *Nilsson* in K. svenska VetensAkad. Handl., ny följd 24, 14: 28 (1892); Durand & Schinz, Consp. Fl. Afr. 5: 421 (1895); N. E. Br. in F.C. 7: 5 (1897);

Malme apud Norlindh & Weimarck, in Bot. Notiser 1932: 11 (1932), and in Ark. Bot. 24A, 5: 4 (1932); Hepper in Kew Bull. 21: 424 (1968); Lewis in Fl. Cameroun 22: 50 (1981). Type: Transvaal, Houtbosch, *Rehmann* 5764 (Z, holo.!; Z, iso.!; BM, iso.!).

X. rigidescens Welw. ex Rendle, Cat. Afr. Pl. Welw. 2,1: 67 (1899). Type: Angola, Lopollo, *Welwitsch* 2474 (BM, holo.!).

X. dispar N.E. Br. in F.T.A. 8: 12 (1902). Type: Zimbabwe, near Salisbury [Harare], *Cecil* 152 (K, holo.!).

Tall perennials forming large tufts; rhizome hard, thick; roots thin. *Leaves* rather coarse, linear, up to 650 × 3–4 mm, tapering to an acute apex; persistent wide leaf-bases up to 150 mm long, shining orange-brown below; cataphylls resembling leaf-bases, caudicled. *Scape* up to 1 m tall, terete to somewhat compressed. *Spikes* subspherical, 10 × 12 mm,

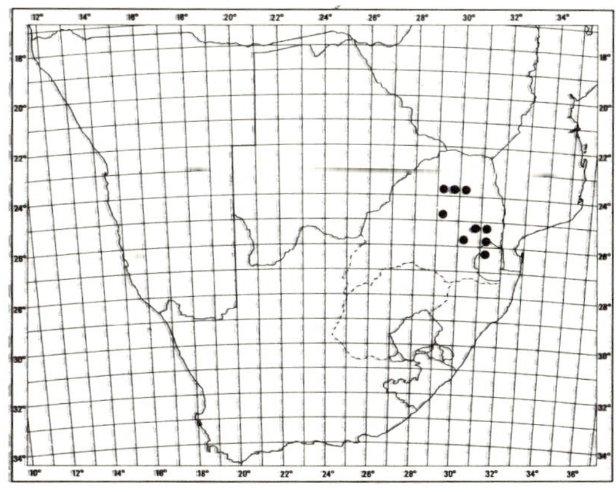

MAP 7.— **Xyris rehmannii**

aristate bracts and lateral sepals give it a prickly appearance. *Lower bracts* large, broadly elliptic, apiculate, stiff, shiny dark brown; fertile bracts smaller, and light brown, often with a central grey patch below an aristate apex. *Lateral sepals* exserted from bracts, curved, keels hard, dark, margin shortly and regularly ciliate, apex bearing a short bent bristle. *Capsule* narrowly ovoid, c. 3 mm long; seeds ovoid. Plate 1: 3.

Recorded from the warmer parts of the Transvaal and Swaziland; also in tropical Africa in Cameroun, Angola and Zimbabwe; in swamps and beside streams. Map 7.

Vouchers: *Burtt Davy* in PRE 1606; *Compton* 26488, 31379; *McCallum* 649; 137; *Mogg* 14706.

Species insufficiently known

Xyris filiformis sensu N. E. Br. in F.C. 7,1: 7 (1897), non Lam. (1791), and in F.T.A. 8,1: 19 (1901), in synon. *X. stramineae* Nilsson, sed falso. Described from a specimen reportedly collected at Apies (sic) river, Transvaal by J. Burke.

Excluded species

Xyris decipiens N. E. Br. in F.C. 7,1:3 (1897). Type: Angola, *Curror* s.n. (K, holo.!). Mistakenly believed to have been collected in our area.

Xyris multicaulis N. E. Br. A specimen from Aapies river, *Burke* s.n. (K!), determined by Malme as *X. multicaulis* N. E. Br., a species described from Malawi [Nyasaland], is a variant of *Xyris capensis* Thunb. of unusual habit.

ERIOCAULACEAE

by A. A. OBERMEYER

Hygrophytic tufted annuals or perennials forming a basal leaf rosette or rarely a leafy stem. *Roots* many. *Leaves* radical, spirally arranged, numerous, linear. *Inflorescence* a capitulum on a naked peduncle, surrounded at the base by a well developed sheath, which in the early stages envelopes both capitulum and peduncle; involucral bracts somewhat to much larger than floral bracts. *Flowers* very small, numerous, crowded, unisexual, the capitula bearing male and female flowers variously arranged on the receptacle; occasionally some capitula with male flowers only. *Female flowers* with 3, or rarely 2 free sepals; petals 3, 2 or 0, free or partly fused; ovary 3–2-locular, superior; style simple with 3–2 long, filiform simple or bifid branches; appendages often present; ovule solitary and pendulous in each locule. *Male flowers* with 3–2 sepals, free or fused into an oblique tube; petals 3–2, placed on a stipe, often with a dark gland on inner surface near apex; stamens 6–4, the filaments free, anthers bilocular, dorsifixed, introrse, black or rarely white; rudimentary pistil in centre. *Capsule* loculicidal, 3–2-lobed; seed ovoid, minute, endospermous.

Genera 13, species c. 500, tropical and subtropical regions, predominantly South American, rare in colder climates. Genera 2 in Southern Africa.

The family description given here is based mainly on the 2 Southern African genera. The South American genera and species exhibit far more specialization; some are suffrutescent with leafy stems and some produce corymbose inflorescences; in others the involucral bracts are enlarged and their heads therefore resemble those of some Asteraceae.

1 Stamens 6–4; scapes and leaves glabrous; petals of female flowers free or 0; style with 3–2 filiform stigmatic branches, without appendages; peduncular sheath breaking up irregularly at apex .... 1. **Eriocaulon**
1 Stamens 3; scapes and leaf-bases glandular-hairy and setose; petals of female flowers united in the middle; style with 3 filiform stigmatic branches and 3 filiform sterile appendages ending in swollen tips; peduncular sheath forming and arum-like spathe at apex .... 2. **Syngonanthus**

828 1. ERIOCAULON

Eriocaulon *L.*, Sp. Pl. 87 (1753); N. E. Br. in F.C. 7: 51 (1897), and in F.T.A. 8: 231 (1901); Ruhl. in Pflanzenreich 4, 30 (Heft 13): 30 (1903); M. Friedrich et al. in F.S.W.A. 159 (1967); R. A. Dyer, Gen. 2: 907 (1976). Type species: *E. decangulare* L.

Tufted perennials or annuals, rarely with a leafy stem. *Roots* many, swollen with a white latticed parenchyma. *Rhizome* in perennial species densely woolly. *Leaves* numerous, rosulate, linear-acuminate, glabrous, their bases with a white latticed parenchyma; in species producing stems leaves densely crowded. *Capitula* globose, at first enveloped in a closed cataphyll which splits obliquely and irregularly when peduncle lengthens; involucral bracts 6–12, usually somewhat larger than floral bracts; receptacle discoid, globose or cylindrical, glabrous or pilose. *Female flowers* sessile or shortly pedicelled; sepals 3–2, free, usually dark and resembling bracts; petals 3–2–0, free, usually narrowly spathulate, whitish, soft, often with a dark gland on inner side below apex, occasionally raised above sepals by a stipe; ovary superior, sessile or on a stipe, 3–2-locular with a solitary pendulous ovule in each locule, with the terminal style ending in 3–2 long stigmatic branches. *Male flowers* with sepals 3–2, obliquely fused at the base, lobed above; petals 3–2, placed on a stipe, with a dark gland below apex in some species; stamens 6–4, biseriate, arising from centre of petals, filaments free, erect, anthers introrse, dorsifixed, black or rarely white; rudimentary pistil present in centre, dark. *Capsule* 3–2-lobed, loculicidal, locules globose; seed ovoid, c. 0,75 mm long, endospermous, epidermis hygroscopic, when enlarged, appearing reticulated, the transverse ribs usually more pronounced than the longitudinal ones, and sometimes white-fringed (*E. ruhlandii* Schinz is an exception; here the longitudinal ribs are more strongly developed).

Species c. 250; distribution as for family. In Southern Africa 12 species, 3 of these apparently endemic, the others widespread but absent from dry interior areas and south-western Cape.

The name *Eriocaulon* (woolly stem) is derived from the Greek and refers to the lanate pubescence on the rhizome of perennial species.

The small annuals are probably wind-distributed. They produce much seed, possibly parthenogenetically; there are usually few staminate flowers. They appear wherever there is an open wet space, whether temporary or permanent, dying when the habitat dries up or becomes too cold. The perennial species inhabit permanently wet, usually montane habitats in fresh running water or are immersed aquatics with only the capitula emergent.

Key to species (based mainly on female flowers)

1 Leaves closely arranged on an elongated stem; a submerged aquatic with only the capitula exserted . 1. *E. setaceum*
1 Leaves in basal rosettes; stems suppressed; small marsh plants, ruderals or rarely aquatics:
  2 Annuals, small, fast growing pioneers, dying when the habitat dries up; capitula without coarse white setae, predominantly female, producing much seed; receptacle cylindrical:
    3 Anthers white; sepals 3 or 2; petals 0 . . . . 2. *E. cinereum*
    3 Anthers black; sepals and petals various:
      4 Petals absent, rarely 1, 2 or 3, much reduced; plants c. 100–200 mm tall; usually submerged except for black capitula; testa of seed with transverse, white, fringed ridges . . . . 9. *E. hydrophilum*
      4 Petals present; plants c. 30–100 mm tall; leaf rosettes not submerged; testa of seed with reticulations, not fringed:
        5 Petals with a ciliate margin and a black apical gland; testa of seed with c. 8 longitudinal ridges . . . . 8. *E. ruhlandii*
        5 Petals with a glabrous margin and without a black apical gland; testa of seed reticulate:
          6 Sepals 2; petals 2 . . . . 3. *E. angustisepalum*
          6 Sepals 3 or 2; petals 3:
            7 Sepals 2; petals 3; sepals boat-shaped with a convex, distended papillate keel, protruding laterally from floral bract when in fruit . . . . 7. *E. maculatum*
            7 Sepals 3; petals 3:
              8 Sepals with a smooth margin:
                9 Capitula dark, except involucral bracts; bracts and sepals broadly ovate, obtuse; receptacle usually quite glabrous . . . . 4. *E. abyssinicum*
                9 Capitula whitish with the bracts and sepals linear-aristate, giving it a bristly appearance; receptacle with few to many silky hairs . . . . 5. *E. welwitschii*
              8 Sepals with a fimbriate margin:
                10 Sepals linear, flat, white, with long fine hairs along margin . . . . 5. *E. welwitschii*
                10 Sepals boat-shaped, deeply concave, dark, with margin pale, membranous, bearing some short coarse setae . . . . 6. *E. gilgianum*
  2 Perennials, more robust plants; capitula sparsely to densely covered with short, coarse, white setae; receptacle globose to discoid:
    11 Anthers black; leaves aerial:
      12 Plants solitary, 60–100 mm tall; capitula many, dark, globose, c. 5 mm in diam.; white setae on bracts, sepals and petals usually few; testa of seed with white, transversely fringed ribs . . . . 10. *E. transvaalicum*
      12 Plants rhizomatous, 100–500 mm tall; capitula few, globose to semi-discoid, c. 10 mm in diam., densely covered with short white setae; testa of seed reticulate . . . . 11. *E. dregei*
    11 Anthers white; aquatic with submerged leaves, often viviparous and stoloniferous . . . . 12. *E. africanum*

1. **Eriocaulon setaceum** *L.*, Sp. Pl. 87 (1753); Ruhl. in Pflanzenreich 4, 30: 89, t. 9 (1903); Meikle in F.W.T.A. edn 2, 3: 62, fig. 337/9 (1968). Type from India (LINN 105.5).

E. melanocephalum Kunth, Enum. Pl. 3: 549 (1841); Ruhl., l.c. 89 (1903). Type: Brazil, St. Paulo, *Sellow* (K, ex herb. Kunth, holo.).

E. bifistulosum Van Heurck & Müll. Arg. in Van Heurck, Obs. Bot. 105 (1870); N. E. Br. in F.T.A. 8: 239 (1901); Ruhl. in Pflanzenreich 4, 30: 90 (1903); H. Hess in Ber. schweiz. bot. Ges. 65: 130, t. 7: figs 3, 4, 5, p. 127: fig. 3 (1955). Type: Tropical Africa, Niger Territory, *Barter* 1021 (K, holo.).

Rooted aquatic. *Stems* elongated, densely leafy, floating near surface of water. *Leaves* numerous, filiform, 20–100 mm long, weak. *Capitula* numerous at apex of stem, emergent on long peduncles, sheathed in lower half; involucral bracts broadly ovate, often torn lengthwise; floral bracts ovate, concave, bearing a few minute white setae; receptacle glabrous. *Female flowers* with 3 equal, obovate, concave sepals; petals 3, obovate to spathulate, enclosed by sepals; ovary 3-locular. *Male flowers* pedicellate; sepals forming an obliquely funnel-shaped, 3-lobed tube; petals stipitate, minute; stamens 6, anthers black. *Capsule* 3-lobed; seed ellipsoid. Fig. 2: 1.

Recorded from Botswana. Widespread in tropical and subtropical Africa, Asia, Australia and America. Map. 8.

Voucher: *Biggs* M613; *P. Smith* 631.

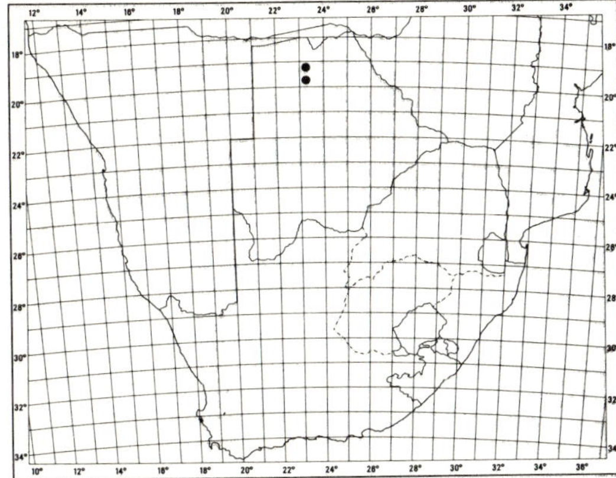

MAP 8.— **Eriocaulon setaceum**

2. Eriocaulon cinereum *R. Br.*, Prodr. 254 (1810); Meikle in F.W.T.A. edn 2,3: 63 (1968). Type: Australia, *R. Brown* (BM, holo.).

Eriocaulon amboense Schinz in Bull. Herb. Boissier 4, App. 3: 35 (1896); N. E. Br. in F.T.A. 8: 258 (1900); Ruhl. in Pflanzenreich 4, 30: 112 (1903); H. Hess in Ber. schweiz. bot. Ges. 65: 176, t. 9: fig. 3, p. 160: fig. 1 (1955); M. Friedrich et al. in F.S.W.A. 159: 1 (1967). Type: South West Africa/Namibia, Ovambo, Uashitenga near Olukonda, *Schinz* 859 (Z, holo.; K).

Small dome-shaped annuals 50–100 mm tall. *Leaves* rosulate, narrowly linear, attenuated into a fine point 10–30 mm long. *Capitula* globose, 3–4 mm in diam., pale at first, grey with age, on peduncles up to 90 mm long; involucral and floral bracts small, ovate; receptacle conical, glabrous. *Female flowers* pedicelled; sepals 3–2, narrowly linear with scattered long white hairs mostly near the base or glabrous; petals 0; ovary on a short stipe, 3-locular; style long, stigmas 3. *Male flowers* typical; stamens 6, bearing white anthers. *Capsule* 3-lobed; seed ellipsoid, smooth. Fig. 2: 2.

Recorded from South West Africa/Namibia, Botswana and from Zimbabwe, Angola, other parts of tropical Africa, Australia, tropical Asia, Japan and America; in seasonally flooded areas. Map 9.

Vouchers: *Giess* 9943; *Merxmüller* 2134; *Smith* 2022.

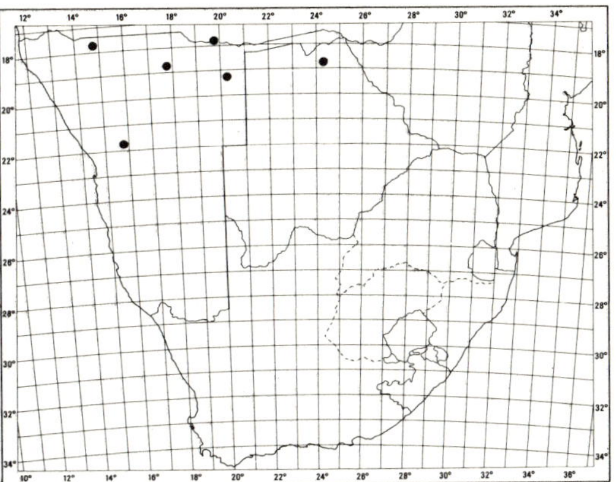

MAP 9.— **Eriocaulon cinereum**

3. Eriocaulon angustisepalum *H. Hess* in Ber. schweiz. bot. Ges. 65: 170, t. 9: figs 6 & 7, p. 160: figs 7 & 8 (1955); M. Friedrich et al. in F.S.W.A. 159: 2 (1967). Type: Angola, Guanhama, road verges 45 km S. of Cassinga, *H. Hess* 52/2004 (ZT, holo.).

Dome-shaped annuals 15–70 mm tall. *Leaves* few, short, linear-acuminate. *Capitula* many, globose, c. 2 mm in diam., black, shiny, on peduncles 15–70 mm long; involucral bracts obtuse, margin erose; floral bracts slightly smaller; receptacle conical, glabrous. *Female flowers* with 2 half-folded, convex, narrow, aristate sepals; petals 2, linear, erect, obtuse; ovary 2-locular, sessile; style with 2 stigmatic branches. *Male flowers* with 2 fused sepals; petals 0; stamens 4, black. *Capsule* bilobed with 2 globose locules; seed minutely reticulate and muricate. Fig. 3: 3.

A pioneer species recorded from Transvaal and Angola; in sandy, marshy places, in seepage areas along road verges or wet rocky outcrops. Map 10.

Vouchers: *Hilliard & Burtt* 9895B; *Strey* 2833, *Venter* 4358a.

4. Eriocaulon abyssinicum *Hochst.* in Flora 28: 341 (1845); N. E. Br. in F.C. 7: 53 (1897), and in F.T.A. 8: 257 (1901); Ruhl. in Pflanzenreich 4, 30 (Heft 13): 282 (1903); H. Hess in Ber. schweiz. bot. Ges. 65: 165, t. 9: fig. 8, p. 160: figs 2 & 3 (1955); Meikle in F.W.T.A. edn 2, 3: 63 (1968). Type: Ethiopia, Shire Province, *Schimper* 1944 (K; G).

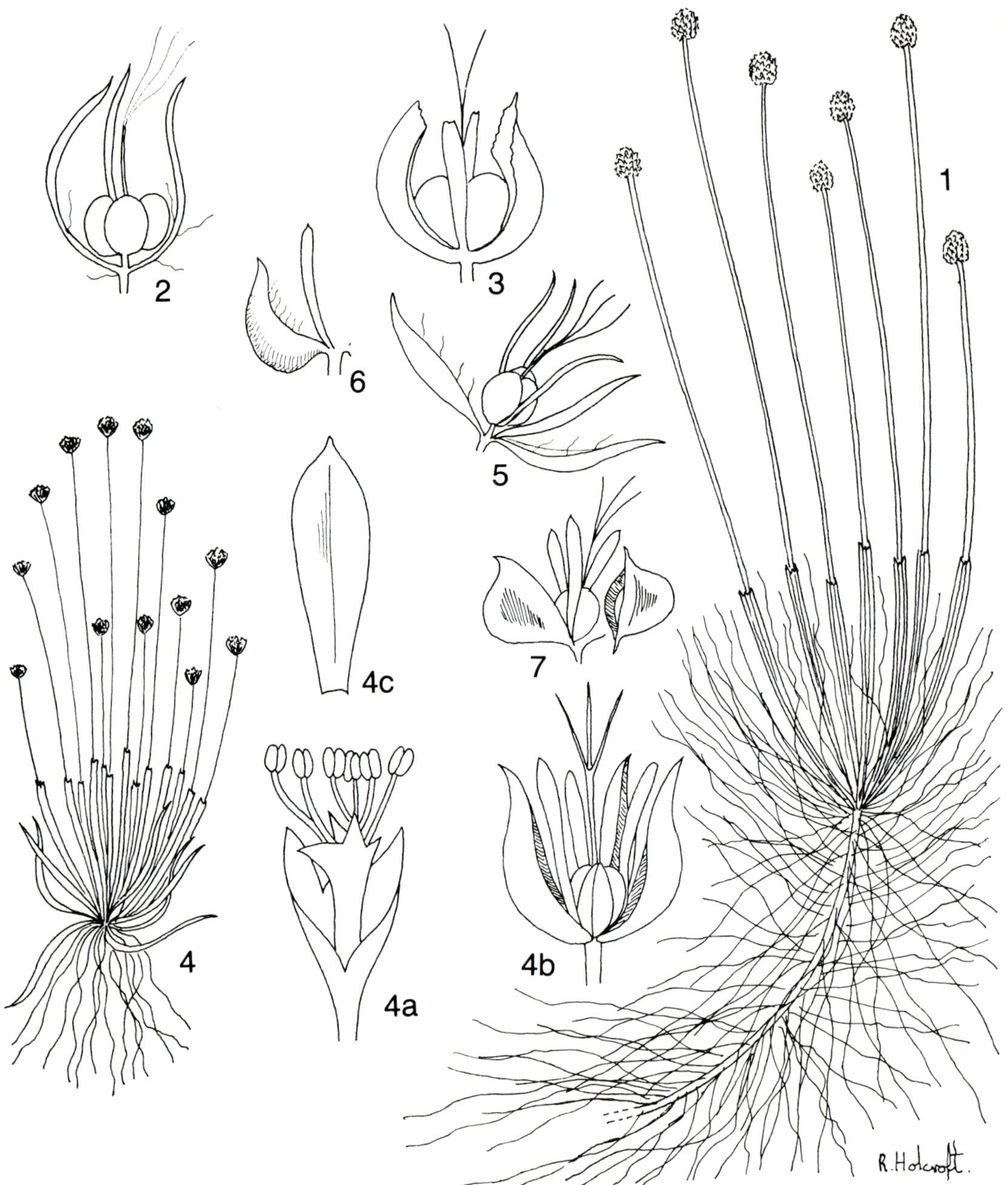

FIG. 2.—1, **Eriocaulon setaceum**, habit, × 0,7 (*Biggs* M613). 2, **E. cinereum**, female flower, × 7 (*Merxmüller* 2134). 3, **E. angustisepalum**, female flower, × 50 (*Venter* 4358A). 4, **E. abyssinicum**, habit, × 2; 4a, male flower, × 20; 4b, female flower, × 18; 4c, floral bract, × 9 (*Killick* 1888). 5, **E. welwitschii**, female flower, × 10 (*Dinter* 7220A). 6. **E. gilgianum**, sepal and petal of female flower, × 7 (*Edwards* 2496b). 7, **E. maculatum**, female flower, × 10 (*Mauve & Venter* 5060).

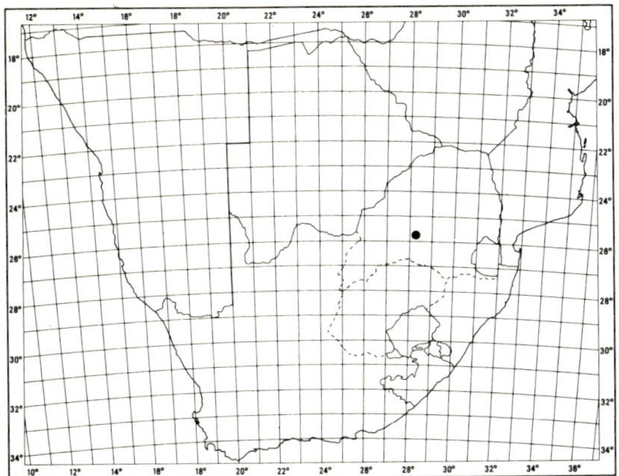

MAP 10.—**Eriocaulon angustisepalum**

Small tufted, dome-shaped annuals 20–100 mm tall. *Leaves* few, linear-acuminate. *Capitula* oblong-globose, c. 3 mm long, dull greyish black, on peduncles up to 45 (–100) mm long; involucral bracts c. 10, pale, about half as long as capitulum, reflexed with age; floral bracts smaller, dark; receptacle conical, glabrous or rarely with a few stray hairs. *Female flowers* on short pedicels; sepals 3, ovate, concave, half-folded, dark; petals linear, acute, glabrous; ovary sessile, 3-locular; style typical. *Male flowers*: sepals 3, obliquely fused below; petal-lobes minute with an apical black gland; stamens 6, with black anthers. *Capsule* trilobed, locules globose; seed ellipsoid, smooth, shiny, but reticulate and muricate under 400× magnification.

Widespread pioneer annual found all over Africa; in Southern Africa recorded from northern South West Africa/Namibia, Transvaal, Orange Free State, Swaziland, Natal and Transkei; in wet places at low or high altitudes.

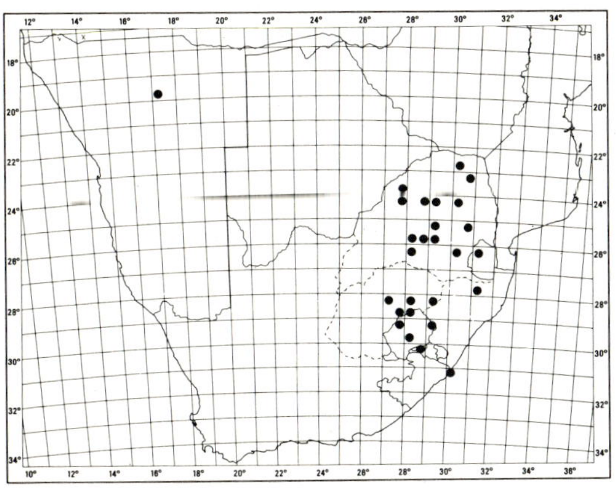

MAP 11.—**Eriocaulon abyssinicum**

Often found growing together with other annual *Eriocaulon* species and *Xyris capensis* seedlings. Map 11.

Vouchers: *Acocks* 22080; *Culverwell* 0674; *Dieterlen* 777; *Killick* 1888; *Leistner, Oliver & Vorster* 218, 313; *Muller* 1881; *Schmitz* 7618; *Van der Schijff* 2844.

5. **Eriocaulon welwitschii** *Rendle*, Cat. Afr. Pl. Welw. 2: 97 (1899); N. E. Br. in F.T.A. 7: 249 (1901); Ruhl. in Pflanzenreich 4, 30 (Heft 13): 102, t. 13 (1903); H. Hess in Ber. schweiz. bot. Ges. 65: 270 (1955). Type: Angola, Pungo Andongo, *Welwitsch 2441* (BM).

E. welwitschii var. *pygmaeum* Rendle, l.c. 98. Type: Angola, Huilla, near Lopollo, *Welwitsch 2444* (BM, holo.; K, iso.).

E. aristatum H. Hess in Ber. schweiz. bot. Ges. 65: 163, 271, t. 9: fig. 5, p. 160: figs 11 & 12 (1955). Type: Angola, Humpata district, *H. Hess 52/1755a* (ZT, holo.).

Small dome-shaped annuals up to 40 mm tall. *Leaves* many, linear-acuminate, 10–20 mm long. *Capitula* oblong-globose, c. 2,5 mm in diam., numerous, on filiform peduncles; involucral bracts c. 10, narrowly ovate-aristate, white, shiny, about half as long as capitulum, recurved when old; floral bracts resembling involucral bracts but somewhat smaller; receptacle conical to cylindrical, with soft, long hairs. *Female flower*: sepals 3, narrowly linear, white, margin glabrous or fimbriate; petals 3, filiform, slightly exceeding sepals; ovary 3-locular, sessile to shortly stipitate, style with 3 long stigmas. *Male flower*: typical, stamens 6, black. *Capsule* 3-lobed; seeds with the testa reticulate. Fig. 2: 5.

Recorded from South West Africa/Namibia, and Botswana, as well as Angola and Zimbabwe and further north; in seasonally flooded areas. Map 12.

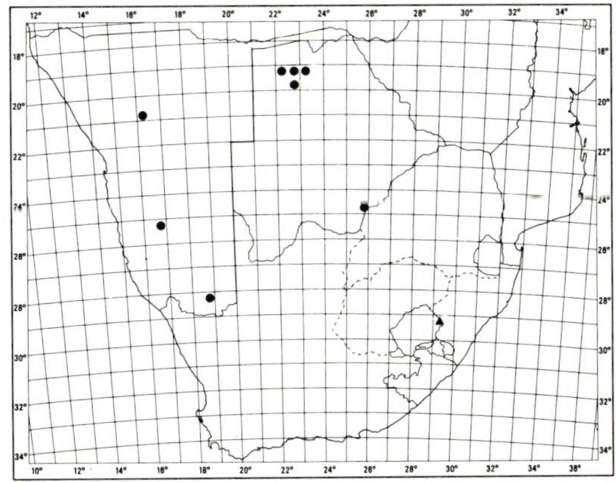

MAP 12.—● **Eriocaulon welwitschii**
▲ **Eriocaulon gilgianum**

Vouchers: *Dinter* 7220; *Smith* 1812; *Volk* 1806.

E. aristatum H. Hess has sepals with a fimbriate margin whereas in *E. welwitschii* they are described as entire, glabrous. However, *Dinter* 7220, a gathering consisting of many plants, was seen to possess either fimbriate sepals or entire ones. No other important differences were observed.

6. **Eriocaulon gilgianum** *Ruhl.* in Bot. Jb. 27: 84 (1899), and in Pflanzenreich 4, 30 (Heft 13): 99, t. 13 (1903); N. E. Br. in F.T.A. 8: 257 (1901); H. Hess in Ber. schweiz. bot. Ges. 65: 158, t. 8: figs 11,12, p. 160: figs 4, 5, 6, 12; p. 270 (1955). Type: Angola, *Antunes* 168 (B, PRE, photo.!).

Dome-shaped annuals up to 130 mm tall. *Leaves* linear-acuminate, c. 10 mm long. *Capitula* oblong-globose, 2–3 mm in diam., dark grey on peduncles 20–100 mm long; involucral bracts c. 10, large, pale, acute, floral bracts smaller, dark, but much larger than sepals; receptacle conical with some long hairs. *Female flower*: sepals 3, c. 1–2 mm long, conchiform, not keeled, enclosing rest of flower, ovate-acuminate, blackish, margin whitish, bearing some coarse setae; dorsal sepal somewhat smaller; petals 3, narrowly spathulate, somewhat longer than sepals; ovary 3-locular; style short with 2 longer stigmas. *Male flower* typical; anthers black. *Capsule* 3-lobed, seeds ovoid, minutely reticulate and muricate. Fig. 2: 6.

Only once recorded from Southern Africa, namely from the Natal Drakensberg, but widespread in Angola and Zimbabwe (often collected at Victoria Falls in rain forest area). Map 12.

Voucher: *Edwards* 2496B (mixed with *E. hydrophilum*).

The Natal plants, coming from a colder climate, are much smaller than those from the main distribution area in the tropics north of our borders. Further collections may shed more light on this taxon.

7. **Eriocaulon maculatum** *Schinz* in Bull. Herb. Boissier sér. 2, 6: 709 (1906). Type: Transvaal, Blouberg, *Schlechter* 4651 (Z, holo.; PRE!; SAM!).

Small annuals up to 100 mm tall. *Leaves* rosulate, linear, c. 15 mm long, 1–2 mm broad, apex acute, soft. *Capitula* usually several, oblong-globose, c. 5 mm in diam., pale to dark brown on erect, long, thin peduncles 100–160 mm long; involucral and floral bracts ovate, obtuse; receptacle cylindrical, with many long, silky hairs. *Female flower*: sepals 2, c. 2 mm long, lateral, boat-shaped with a broad curved keel consisting of large bulbous cells, with a dark central area, the sides rounded above, attenuated below, enveloping the ripe seed at maturity; petals 3, unequal, narrowly linear, exserted. *Ovary* on a stipe, 3-locular; style with 3 stigmas. *Capsule* 3-lobed; seeds ovoid, the testa transversely ribbed, the ribs white-fringed. Fig. 2: 7. Plate 2: 1.

Recorded from N. Transvaal to Zimbabwe, Zambia and Malawi; in moist sandy areas along streams, marshes or temporary wet places. Map 13.

Voucher: *Mauve & Venter* 5060.

This species closely resembles the tropical species *E. buchananii* Ruhl., but the specimens of our species studied lacked the reduced third posterior sepal in the female flower, and there are differences in the surface of the testa.

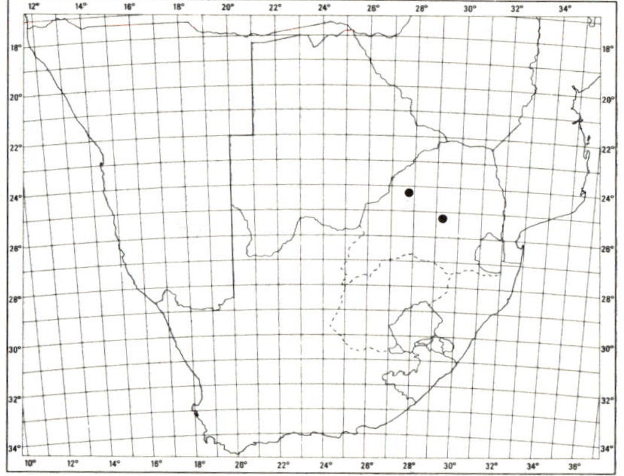

MAP 13.—**Eriocaulon maculatum**

8. **Eriocaulon ruhlandii** *Schinz* in Bull. Herb. Boissier sér. 2, 6: 710 (1906); Ross in Fl. Natal 116 (1972). Type: Natal, near Clairmont, *Schlechter* 2955 (Z, holo.; K!; PRE!; NH!; GRA!).

Annuals up to c. 0,16 m. *Capitula* globose, light to dark brown and shiny, c. 4 mm in diam., with ovate involucral and floral bracts; receptacle cylindrical with long fine silky hairs. *Female flower*: sepals 2, linear, c. 2–3 mm long, dark, glabrous; petals 3, narrowly spathulate, white with a black spot at apex, margin with long cilia; ovary sessile, 3-locular, style with 3 stigmas. *Male flower*: sepals 2, free, broadly linear-acuminate, dark; petals white with an apical gland; stamens 4, anthers black. *Capsule* 3-locular; testa of seed with c. 8 longitudinal ridges. Fig. 3: 1. Plate 2: 2.

Largely confined to the coastal region of Natal and northern Transkei; in wet areas. Map 14.

Vouchers: *Strey* 6507; *Ward* 8130, 9097.

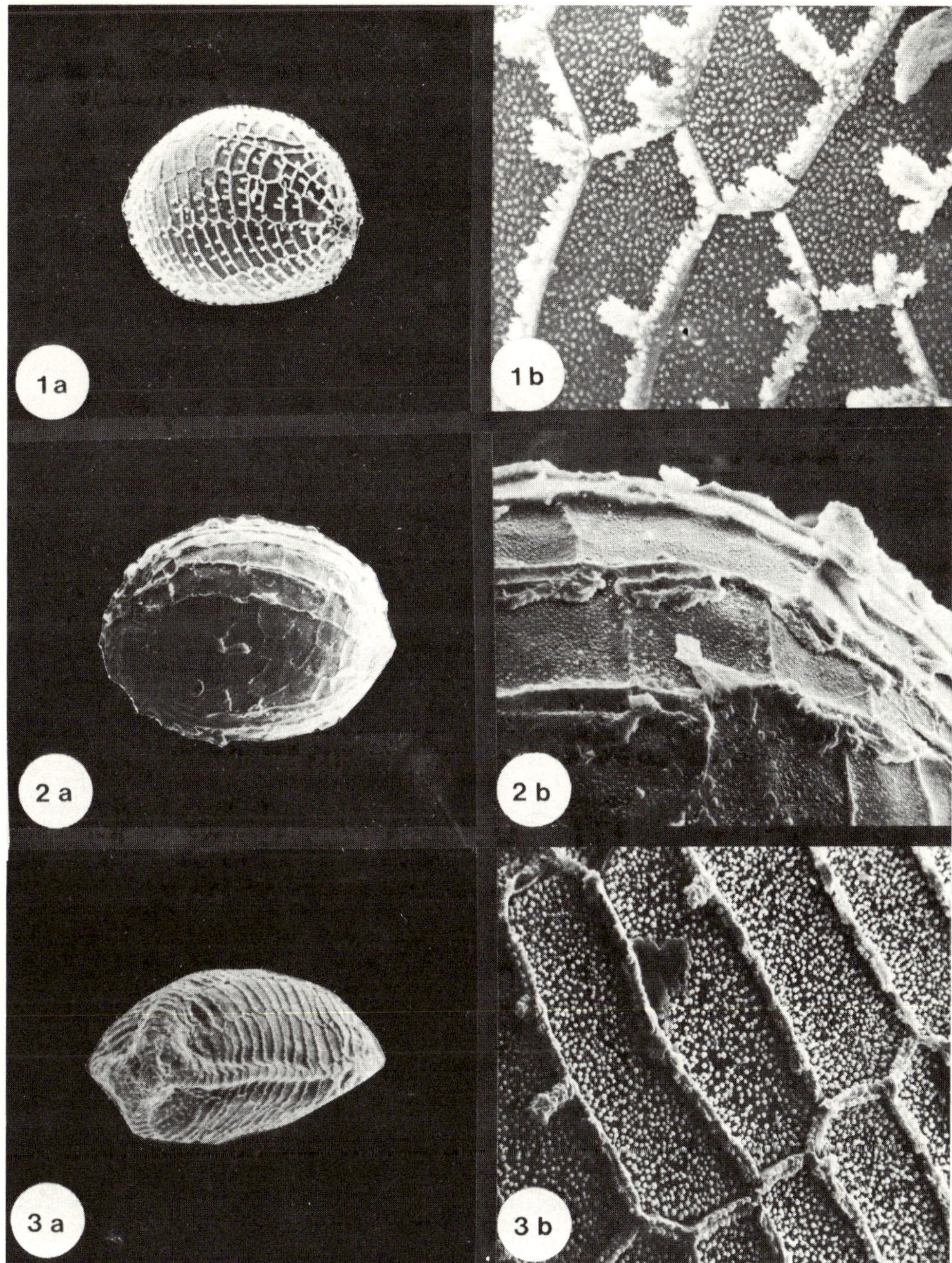

PLATE 2.—Seeds of **Eriocaulon** species: 1, **E. maculatum**: 1a, × 120; 1b, × 1200 (*Mauve & Venter* 5060). 2, **E. ruhlandii**: 2a, × 120; 2b, × 1200 (*Schlechter* 2955). 3, **E. hydrophilum**: 3a, × 120; 3b, × 1200 (*Hilliard & Burtt* 9026).

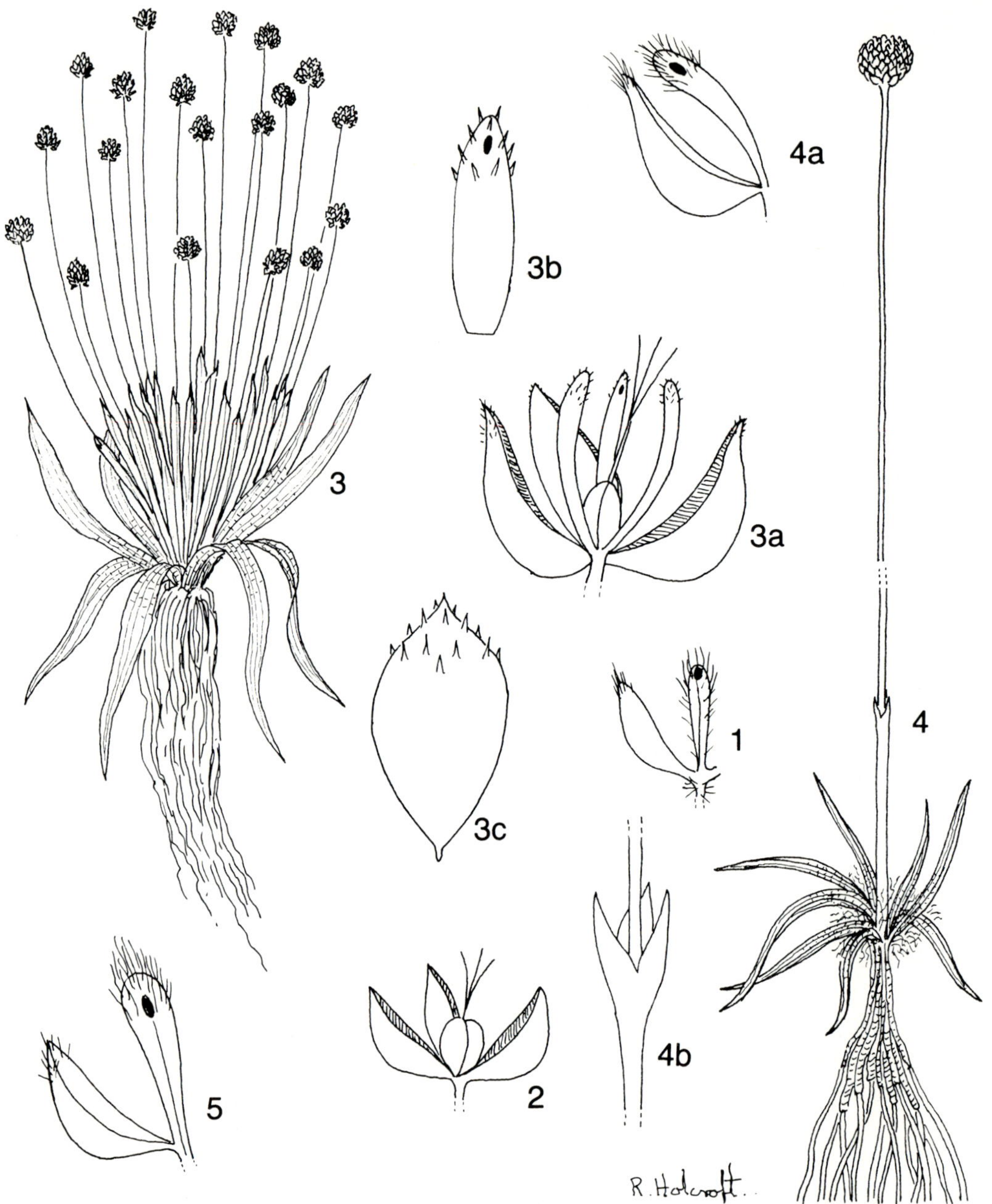

Fig. 3.—1, **Eriocaulon ruhlandii**, sepal and petal, × 5 (*Ward* 2868). 2, **E. hydrophilum**, female flower, × 6 (*Burtt & Hilliard* 9026). 3, **E. transvaalicum**, habit, × 0,5; 3a, female flower, × 12; 3b, petal, × 15; 3c, floral bract, × 18 (*Bosman* sub PRE 37863). 4, **E. dregei** var. **sonderianum**, habit, × 1; 4a, sepal and petal, × 8; 4b, sheath of peduncle, × 2 (*Mauve* 5033). 5, **E. africanum**, sepal and petal, × 8 (*Strey* 7629).

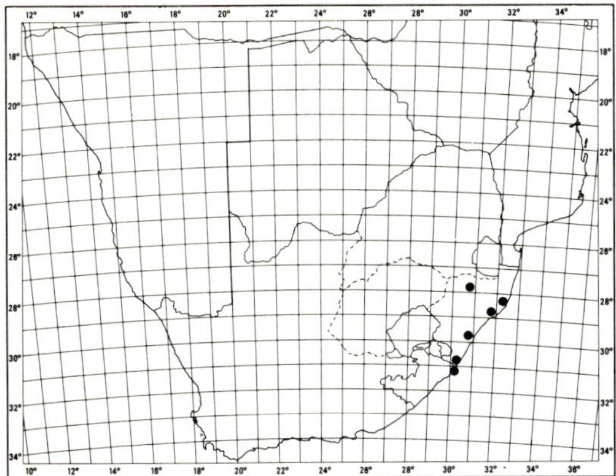

MAP 14.— **Eriocaulon ruhlandii**

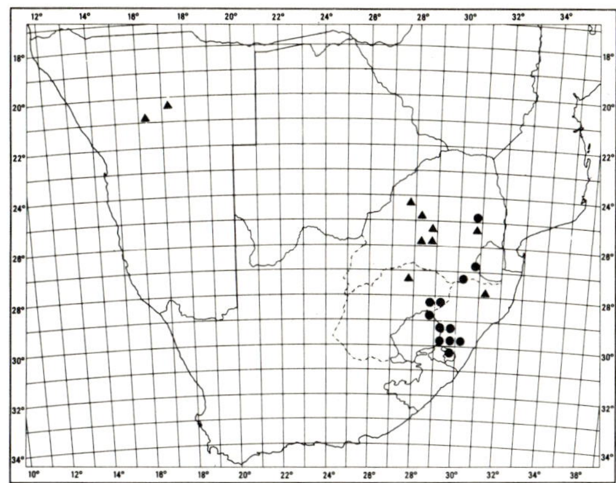

MAP 15.— ● **Eriocaulon hydrophilum**
▲ **Eriocaulon transvaalicum**

9. **Eriocaulon hydrophilum** *Markötter* in Ann. Univ. Stellenbosch 8: 10 (1930). Type: Orange Free State, Witzieshoek, Quaqua Mtn, *Thode* 8274 (STE, holo.!).

Annuals, usually submerged with only the heads emergent, c. 100 mm(−260 mm) tall. *Leaves* rosulate, linear-acuminate, with long filiform tips, thin, soft. *Capitula* small, depressed-globose, 3–5 mm in diam., shiny black; peduncles erect, thin; involucral bracts c. 6, ovate; floral bracts smaller, receptacle shortly cylindrical, glabrous or with a few long soft hairs. *Female flower*: sepals 3, c. 2,5 mm long, equal, half-folded, enclosing inside of flower, keel convex; petals 0, rarely 1, 2 or 3, small, linear; ovary 3-lobed; style with 3 stigmas. *Male flower*: few towards apex; sepals linear, fused below to form an oblique 3-lobed tube; petals 3, minute; stamens 6, black. *Capsule* 3-lobed; seeds with white transverse, fringed ridges. Fig. 3: 2. Plate 2: 3.

A montane species recorded from the eastern Transvaal, eastern Orange Free State, Lesotho and Natal on the Drakensberg complex at altitudes up to 2 000 m. Map 15.

Vouchers: *Beverley & Hoener* 526; *Devenish* 1701; *Hilliard & Burtt* 9026; *Moss* 16429; *Stam* 424.

10. **Eriocaulon transvaalicum** *N. E. Br.* in F.C. 7: 54 (1897); Ruhl. in Pflanzenreich 4, 30 (Heft 13): 81 (1903); H. Hess in Ber. schweiz. bot. Ges. 65: 155 (1955). Type: Transvaal, Bosveld near Boekenhoutskloof, *Rehmann* 4787 (K, holo.).

E. tofieldiifolium Schinz in Bull. Herb. Boissier sér 2,1: 779 (1901); H. Hess, l. c. 266, t. 5 on p. 265 (1955); M. Friedrich et al., in F.S.W.A. 159: 2 (1967). Type: South

West Africa/Namibia, Waterberg, *Dinter* 378 (Z, holo.!, PRE, photo.!).

Fairly robust annuals or perennials up to 0,16 m tall, with many erect, black heads. *Leaves* broadly linear, acute, c. 30 × 5 mm, widening at base. *Capitula* globose, c. 5 mm in diam., shiny black, bearded with a few to many short stiff, white setae; peduncles firm, up to c. 0,16 m long; basal sheaths about as long as leaves, loose; involucral bracts c. 10, somewhat lighter in colour, reflexed with age; floral bracts ovate, acute; receptacle bearing fine, long, white hairs. *Female flower*: sepals 3, c. 2 mm long, the lateral half folded with a convex, acute keel, attenuated below, inner sepal reduced, flat; petals raised on a stipe, linear, widened and dark above, obtuse, exserted; ovary 3-lobed, sessile, style and stigmas typical. *Male flower*: typical, anthers black. *Capsule* 3-lobed; seeds pale yellow with transverse ribs somewhat fringed. Fig. 3: 3.

Recorded from South West Africa/Namibia, Transvaal, Orange Free State, Natal and further north to tropical Africa. Map 15.

Vouchers: *Boss* in TRV 35083; *Repton* 3572, *Van der Schijff* 2131.

11. **Eriocaulon dregei** *Hochst.* in Flora 28: 341 (1845); N. E. Br. in F.C. 7: 55 (1897). Type: Eastern Cape, *Drège* 4101 (K!).

Caespitose perennials forming colonies, compact hard rhizome covered with .white soft tomentum of multicellular hairs. *Leaves* rosulate, linear-acuminate, up to 250 × 8 mm, but usually much smaller, apex acute, obtuse or drawn out into a long fine point. *Capitula*

1–3 per rosette on long thin peduncles, globose, c. 10 mm in diam., usually densely covered by a tomentum of hard, short, white setae; monoecious with male flowers predominating or some capitula entirely male; involucral bracts glabrous, usually reflexed, oblong, pale; floral bracts dark with white setae near apex; receptacle glabrous to villous. *Female flower*: sepals 3, black, boatshaped, 2–4 mm long, with a convex, occasionally toothed keel, apex acute, tufted with white setae, margin fimbriate; petals 3, spathulate, pale, densely setose and with a black gland; ovary 3-lobed. *Male flower*: sepals 3, dark, setose at apex; petals 3, unequal in size, white, setose above with a black gland; stamens 6, anthers black. *Capsule* 3-lobed; seeds smooth, testa with a rectangular network; ripe seeds rare.

Widespread in the eastern parts of Southern Africa from the eastern Cape to Transkei, Natal, eastern Orange Free State, Swaziland, and Transvaal; in montane areas around springs.

Two varieties are recognized:

1 Leaves c. 100–250 mm long, linear-acuminate, the apex obtuse, ending in a small knob; larger plants found mainly in the eastern Cape to Natal . . . . (a) var. *dregei*
1 Leaves up to 80 mm long, long acuminate ending in a fine point; smaller plants found mainly in the Transvaal and Natal . . . . (b) var. *sonderianum*

(a) var. **dregei**.

E. dregei Hochst. in Flora 28: 341 (1845); N. E. Br. in F.C. 7: 55 (1897).

Plants robust, up to 0,35 m tall. *Leaves* linear, attenuated into an obtuse apex, ending in a small knob, up to 250 × 10 mm, but usually smaller. *Capitula* globose, c. 12 mm in diam. when bearing male and female flowers, smaller and more discoid when male flowers only are present. *Flowers* 2–4 mm long.

Common in Natal, also in eastern Cape, Transkei and Transvaal; usually montane but also recorded from near the coast. Map 16.

Vouchers: *Hilliard* 2146; *Kluge* 821; *Pooley* 1759; *Sim* 2828; *Strey* 4467.

(b) var. **sonderianum** *(Koern.) Oberm.*, comb. nov.

E. sonderianum Körn. in Linnaea 27: 669 (1854); N. E. Br. in F.C. 7: 55 (1897). Syntypes: Transvaal, Magaliesberg, *Zeyher* 1731 (K; SAM!); *Burke* s.n. (K).

E. baurii N. E. Br. in F.C. 7: 54 (1897). Type: Transkei, Bazeia Mountain, *Baur* 622 (K, holo.; SAM!).

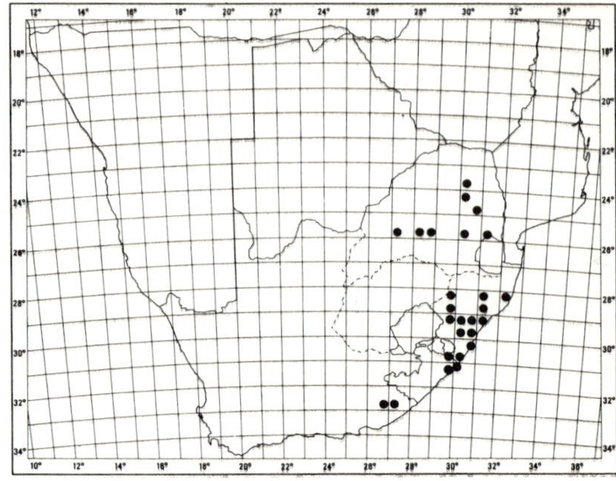

MAP 16.—**Eriocaulon dregei** var. **dregei**

Plants smaller, up to 100 mm tall. *Leaves* linear-acuminate, 50–100 mm long, apex attenuated into a long filiform point. *Capitula* globose to depressed globose, c. 8 mm in diam., bearing male and female flowers (no capitula with only male flowers seen). *Flowers* c. 2 mm long. Fig. 3: 4.

Transvaal, Swaziland, Natal, Lesotho and Transkei; usually montane and found at high altitudes. Map 17.

Vouchers: *Baur* 622; *Beverley & Hoener* 528; *Codd* 6721; *Compton* 20268; *Flanagan* 1863; *Killick* 4178; *Moll* 1432b; *Venter* 690.

Mr R. D. Meikle of Kew suggested that *E. sonderianum* could be a variety of *E. dregei*, and the ample collections available now support this view. *E. dregei* appears to be a polyploid. *E. dregei* var. *sonderianum* is the common variety in the Transvaal; in Natal both varieties occur but in the eastern Cape the larger var. *dregei* becomes common whereas var. *sonderianum* is rare.

At high altitudes, where ice and snow occur in winter, the dwarfed plants of var. *sonderianum* are found.

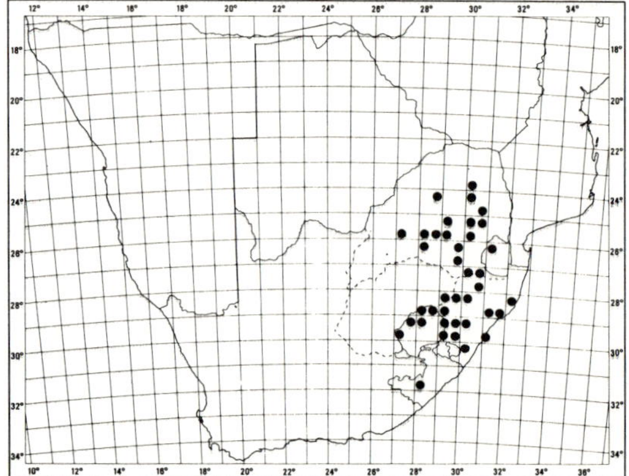

MAP 17.— **Eriocaulon dregei** var. **sonderianum**

The sheath that covers the young inflorescence splits apically into 4 small spreading dry lobes, which is very characteristic of this form. Normally the sheath bursts sideways forming one large, somewhat hooded lobe and 1 or 2 small ones.

12. **Eriocaulon africanum** *Hochst.* in Flora 28: 340 (1845); N. E. Br. in F.C. 7: 56 (1897); H. Hess in Ber. schweiz. bot. Ges. 65: 266 (1955). Type: Natal, Mgeni River near Pietermaritzburg, *Krauss* 375 (K, iso.!).

E. woodii N. E. Br. in F.C. 7: 57 (1897). *E. natalensis* Schinz in Mém. Herb. Boissier No. 10: 76 (1900). Type: near Murchison, *Wood* 3053 (K, holo.!; NH!).

Perennial, often stoloniferous aquatics, submerged except for the emergent capitula, often viviparous. *Leaves* rosulate, linear-acuminate, 40–200 mm long, apex acute, soft. *Capitula* subglobose, c. 10 mm in diam., dark; peduncles long, thin; involucral bracts c. 10, short, rounded, glabrous; floral bracts narrower, black, glabrous; receptacle discoid. *Female flower*: sepals 3, free, boat-shaped, black with sparse white setae; petals 3, narrowly obovate, attenuated below, white, with an apical black gland and setose above; ovary 3-lobed, style short, stigmas 3, long. *Male flower*: predominant, sepals 3, fused below, lobes broad, black, sparsely and irregularly setose; petals 3, obovate, white with an apical black gland and densely setose above; stamens 6, anthers pale yellow. *Capsule* 3-lobed; testa of seed forming an irregular network of 4–6-sided rectangles. Fig. 3: 5.

Recorded from Transvaal, Natal and Transkei and from Zimbabwe and Zambia; rare, usually in flowing water. Map 18.

Vouchers: *Hemm* 216; *Schlechter* 3319; *Strey* 7629; 6856; *Tyson* 2551.

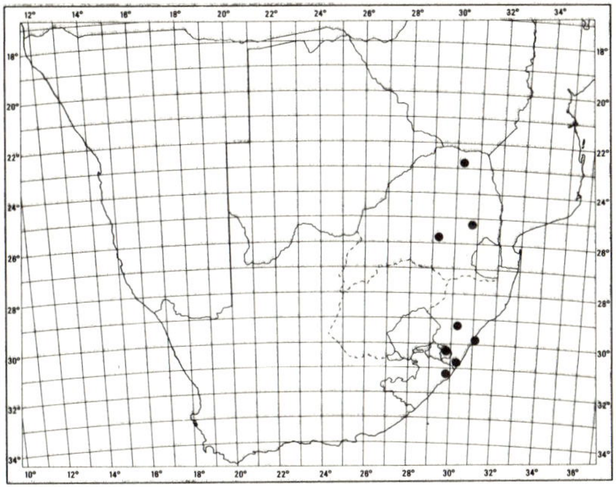

MAP 18.—**Eriocaulon africanum**

832A   2. SYNGONANTHUS

Syngonanthus *Ruhl.* in Urban, Symb. Antill. 1: 487 (1900), and in Pflanzenreich 4, 30 (Heft 13): 242 (1903), and in Natürl. PflFam. edn 2, 15a: 55 (1930); Meikle in F.W.T.A. edn 2, 3: 67 (1968); R.A. Dyer, Gen. 2: 908 (1976). Type species: *S. umbellatum* (Lam.) Ruhl.

Annuals or perennials with a basal rosette of leaves (in Southern African species). *Capitula* on long peduncles sheathed at base. *Female flowers*: sepals 3, free; petals 3, with free claws, fused in middle, with free lobes above; ovary 3-locular, style simple, with 3 filiform, glandular-papillate stigmatic branches alternating with 3 glabrous filiform appendages. *Male flowers*: sepals 3, free; petals 3, connate into a subtruncate tube; stamens 3, anthers white. *Capsule* 3-lobed; seeds with a few longitudinal ribs (in Southern African species).

Species c. 80, mostly from South America with a few in tropical and subtropical Africa, one reaching the Transvaal.

Syngonanthus wahlbergii (*Koern.*) *Ruhl.* in Pflanzenreich 4, 30 (Heft 13): 247 (1903); Meikle in F.W.T.A. edn 2, 3: 67 (1968).

Paepalanthus wahlbergii Koern. in Mart. Fl. Bras. 3, 1: 459 (1863); N. E. Br. in F.C. 7: 69 (1897) and in F.T.A. 8: 263 (1902). Type: Transvaal, Kaapse Hoop, *Wahlberg* (S, holo., PRE, photo.!).

Small compact perennials c. 50–100 (–160) mm high, all parts sparsely to densely covered by patent gland-tipped hairs. *Rhizome* compact, densely lanate from disintegrating leaf-bases; roots long, white spongy. *Leaves* numerous (± 100) in compact rosettes, the upper arched inwards, subulate, c. 12 mm long. *Capitula* numerous, developing in close succession, globose, c. 5 mm in diam.; peduncle erect, thin, glandular-pubescent especially so below capitulum; basal sheath cylindrical, forming an open arum-like spathe at apex. *Flowers* numerous, small, pedicelled on a villous receptacle; involucral bracts more or less similar to floral bracts,

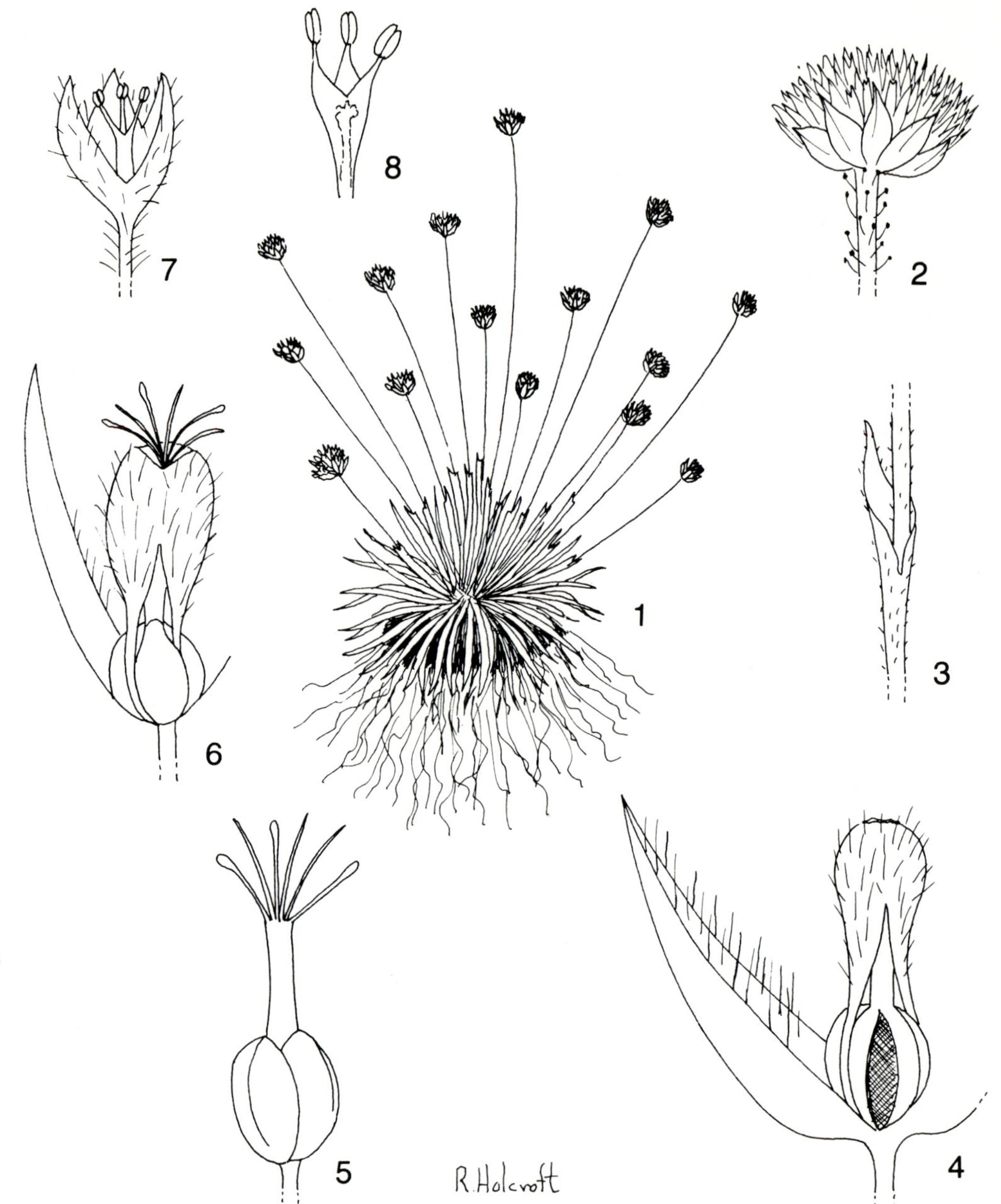

FIG. 4.—**Syngonanthus wahlbergii**: 1, habit, × 1; 2, capitulum, × 5; 3, spathe at base of peduncle, × 7; 4, female flower in fruit, showing empty locule and tepals fused and shrunk above, × 30; 5, gynoecium, × 90; 6, female flower at anthesis, × 30; 7, male flower, × 30; 8, stamens with rudimentary gynoecium inside tube, × 30 (*Mauve & Venter* 5197).

ovate, glabrous, membranous, shiny, light to dark brown. *Female flowers*: sepals 3, ovate, half-folded, acute, fimbriate; petals 3, transparent, villous, free below, fused above, apical lobes short, triangular, soon rolled inwards; ovary 3-locular; style fairly stout, cylindrical, with 3 apical, filiform, glandular stigmatic branches and 3 filiform, glabrous appendages somewhat swollen at the apex. *Male flowers:* sepals 3, ovate; petals 3, on a short stipe, crenulate apically; stamens 3. *Capsule* trilobed; seed ovoid with c. 8 longitudinal ridges. Fig. 4.

Widespread in tropical and subtropical Africa. Recorded from warmer parts of the Transvaal, with a single record from Natal; in running water, the leaf-rosettes submerged. Map 19.

Vouchers: *Du Plessis* 860; *Mauve & Venter* 5175; *Nelson* 294; *Schlechter* 3718.

One of the plants mounted on the type sheet of *Eriocaulon ruhlandii* Schinz (Z), viz. *Schlechter* 2955 from Claremont (Natal: 2930 DD, as 'Clairmont') belongs to *S. wahlbergii* (cf. H. Hess in Ber. schweiz. bot. Ges. 65: 271; 1955). This appears to be the only record so far from Natal. The sheet *Schlechter* 2955 at Kew has no *Syngonanthus* specimens.

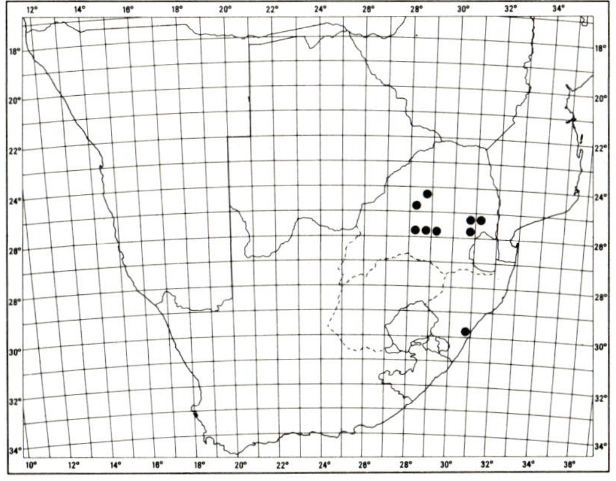

Map 19.—**Syngonanthus wahlbergii**

COMMELINACEAE*

by A. A. OBERMEYER and ROBERT B. FADEN**

Perennials, chamaephytes or annuals, often somewhat succulent; the perennials of diverse habits, sometimes rhizomatous or stoloniferous, very rarely forming a small bulb; roots adventitious, fibrous, thin or swollen. *Leaves* basal and/or cauline, spirally arranged, with a basal, usually closed sheath enveloping the stem, often ciliate at the mouth; blade often petiolate. *Inflorescences* terminal, or terminal and axillary, rarely all axillary, composed of cymes which may be few or numerous and aggregated into thyrses; sometimes subtended by or enclosed in spathaceous bracts. *Flowers* regular or zygomorphic, bisexual or male, occasionally cleistogamous. *Sepals* 3, free or united, persistent. *Petals* 3, free or united, deliquescent. *Stamens* 6, hypogynous or united with corolla, in 2 whorls; in some genera 1–4 modified into staminodes or some suppressed; filaments often bearded; anthers basifixed, dorsifixed or versatile, opening by longitudinal slits or rarely by basal or apical pores. *Ovary* superior, 2–3-locular with 1–many ovules in each locule; style simple, stigma apical, small or capitate. *Fruit* a 2–3-valved capsule, rarely indehiscent; seeds with a dot-like or elongate hilum and a dorsal to lateral (rarely terminal) circular embryotega.

Genera about 50, species about 700, cosmopolitan in warm regions; a few of them widespread weeds. Zebrina pendula *Schnizl.* (= *Tradescantia zebrina* Loud.) has been recorded as a garden escape in southern Natal.

Lit. Brenan in Kew Bull. 7: 179 (1952); Brenan in Kew Bull. 14: 280 (1960); Brenan in Kew Bull. 15: 207 (1961); Brenan in Kew Bull. 19: 63 (1964); Brenan in J. Linn. Soc., Bot. 59: 349 (1966); Brenan in F.W.T.A. edn 2, 3: 22 (1968); Brückner in Bot. Jb. 61, Beibl. 137: 1 (1926); Brückner in Pflanzenfam. edn 2, 15a: 159 (1930); C. B. Cl. in A. DC., Monogr. Phan. 3: 113 (1881); C. B. Cl. in F.C. 7: 7 (1897); C. B. Cl. in F.T.A. 8: 25 (1901); Faden in Agnew, Upland Kenya Wild Flowers 653 (1974); Faden & Suda in Bot. J. Linn. Soc. 81: 301 (1980); Jones & Jopling in Bot. J. Linn. Soc. 65: 129 (1972); Lewis & Tadesse in Kirkia 4: 213 (1964); Lewis in Sida 1: 274 (1964); Morton in J. Linn. Soc., Bot. 55: 507 (1956); Morton in J. Linn. Soc., Bot. 59: 431 (1966); Morton in J. Linn. Soc., Bot. 60: 167 (1967); Tomlinson in Metcalfe, Anatomy of Monocotyledons 3 (1969); Tomlinson in J. Linn. Soc., Bot. 59: 371 (1966); Troll, Beitr. Biol. Pflanzen 36: 326 (1961).

1 Flowers actinomorphic; stamens 6, all fertile:
  2 Inflorescences axillary, perforating the sheaths . . . . 4. **Coleotrype**
  2 Inflorescences terminal or terminal and axillary, never perforating the sheaths:
    3 Flowers white; petals free; cymes regularly paired . . . . 7. **Tradescantia**
    3 Flowers pink or purple to violet; petals united basally into a tube; cymes usually not
      paired . . . . 5. **Cyanotis**
1 Flowers slightly to strongly zygomorphic; stamens 6, usually unequal, all fertile or 3–4
  sterile:
  4 Inflorescences consisting of 1–2 cymes enclosed in or closely subtended by folded or
    obliquely funnelshaped leafy bracts (spathes); staminodes typically with 4(–6)-lobed
    anthers . . . . 1. **Commelina**
  4 Inflorescences not enclosed in bracts, consisting of 1 to many cymes; staminodes lacking or (if present) with 2–3-lobed anthers:
    5 Capsules trivalved; fertile stamens 2; 3 or more filaments bearded . . . . . . 3. **Murdannia**
    5 Capsules bivalved; fertile stamens 3 or 6; all filaments glabrous or 2 bearded and 4
      glabrous:
      6 Fertile stamens 6; ovary and capsule glabrous . . . . 6. **Floscopa**
      6 Fertile stamens 3, staminodes 3; ovary and capsule puberulous . . . . 2. **Aneilema**

---

* The kind cooperation of Prof. J. P. M. Brenan, former Director of the Royal Botanic Gardens, Kew, is gratefully acknowledged.

** Department of Botany, National Museum of Natural History, Smithsonian Institution, Washington, D.C. 20560, U.S.A.

896   **1. COMMELINA**

Commelina *L.*, Sp. Pl. 40 (1753); C. B. Cl. in A. DC., Monogr. Phan. 3: 138 (1881) and in F.C. 7: 8 (1897) and in F.T.A. 8: 33 (1901); R.A. Dyer, Gen. 2: 909 (1976). Type species: *C. communis* L.

Perennial or annual herbs. *Roots* thin or swollen. *Stems* branched, erect to spreading or rarely succulent; when perennial usually dying down in winter to a persistent knobbly crown with long hard roots (chamaephytes); sometimes underground stolons bearing cleistogamous flowers, capsules and seeds are often developed in *C. benghalensis* L. and *C. forskaolii* Vahl. *Leaves* with a sheathing base usually ciliate at mouth. *Flowers* in 1 or 2 1–6-flowered cymes enclosed in a terminal or leaf-opposed, obliquely funnel- to boat-shaped spathe which has either free margins and is dry inside or margins of proximal end are fused and straight and spathe is mucilaginous inside when flowering; upper cyme 1- or rarely 2–3-flowered, usually male (with or without a pistillode), exserted on a long peduncle, in some species absent or represented by a vestigial peduncle, often enclosed within a spathe; lower cyme included, borne on a short peduncle, flowers 2–6, all bisexual or some male, just exserted in turn at anthesis; pedicels recurved inside spathe when in fruit; flowers open in morning for a few hours, soon deliquescent. *Sepals* 3, free or lower 2 fused. *Petals* 3, free; upper 2 flat, long-clawed, blue, white or yellow or related colours, lower one reduced, often colourless. *Stamens* 3, anterior (lower) fertile, with central one usually with a larger anther with a broad connective; 3(–2) posterior (upper) staminodial, reduced, antherodes cruciform or variously deformed. *Ovary* 3(–2)-locular with 2 ventral locules and 1 dorsal locule which may be reduced to absent; ventral locules 2- or 1-ovulate, dorsal locule 1-ovulate or sterile. *Capsule* loculicidal, dehiscent; in some species dorsal locule indehiscent, smooth, or striate-muricate with capsule-wall fused to seed and laterally attached to part of walls of ventral locules, forming a shallow cup. *Seeds* 5(–2), globose, ellipsoid or cylindric-truncate; testa various; hilum linear; embryotega lateral. *Chromosomes*: basic numbers are 15 (in most species), 14, 13, 12 and 11 (Faden & Suda, 1980).

Species about 170 in the warmer countries of the world. In Southern Africa 16 species, widespread.

The genus was named after the Dutch botanists Jan Commelijn (1629–1692) and his nephew Caspar (1667/1668–1731).

Capsule and seeds are important for the identification of species but unfortunately they are often absent from herbarium sheets. In species 1–11 the ventral locules are 2-seeded, but the dorsal locule usually develops only one seed, or it may be sterile and indeshiscent. Species 12–16 have 3 one-seeded locules, but here too the dorsal locule may be indehiscent and often sterile as well. The shape of the seed and its testa are typical for the species.

1 Spathes simply folded, dry inside, proximal end free, convex:
  2 Annuals; spathes subsessile, falcate; cyme solitary, c. 3-flowered, the flowers only just exserted at anthesis, one at a time .... 1. *C. subulata*
  2 Perennials, spathes pedunculate; cymes 2; flower(s) of upper cyme exserted:
    3 Petals blue, rarely white; hygrophilous trailing plants rooting at the nodes:
      4 Leaves narrowly linear, 50–250 mm long, 1–3 mm broad; aquatic or semi-aquatic (S.W.A./Namibia and Botswana) .... 2. *C. fluviatilis*
      4 Leaves broadly linear to ovate, 50–90 mm long, c. 10 mm broad; margin of leaf raised and covered with minute white pustules .... 3. *C. diffusa*
    3 Petals yellow: chamaephytes from drier areas; stems erect or procumbent from a hard knobbly rootstock:
      5 Leaves mesophytic, oblong to linear, 30–60 mm long; stems with normally developed side branches; a widespread variable species .... 4. *C. africana*
      5 Leaves scleroid, wiry, linear, falcate, up to 30 mm long, placed on abbreviated, imbricate side branches; northern Transvaal .... 5. *C. rogersii*
1 Spathes obliquely fused, mucilaginous inside when flowering, proximal side connate, straight (cf. *C. imberbis* and *C. petersii* where margins are connate towards the base only):
  6 Capsule quadrate; ventral locules 2-seeded, constricted between seeds; dorsal locule 1-seeded or often aborted:
    7 Annuals with thin roots (cf. *C. forskaolii* which can be a chamaephyte occasionally):
      8 Leaves (at least upper) sessile and cordate at the base; plants not producing subterranean flowers ....
.... 8. *C. imberbis*

8 Leaves narrowed at the base; both cymes developed; plants usually producing cleistogamous flowers and capsules on subterranean stolons:
   9 Leaves ovate, green, margin smooth, leaf-sheaths with often long red setae at the mouth; petals ink blue; dorsal cell of capsule smooth . . . . 6. *C. benghalensis*
   9 Leaves oblong to linear, usually greyish green, margin strongly undulate, leaf-sheaths with short colourless setae at the mouth; flowers pale sky blue; dorsal cell of capsule tuberculate . . 7. *C. forskaolii*
7 Perennials (usually chamaephytes) with thick roots (slender in *C. petersii*):
  10 Leaves usually linear, soft, smooth; cymes 2; seeds globose with a reticulate testa . . . . 9. *C. eckloniana*
  10 Leaves narrowly ovate-acuminate, rough; upper cyme reduced to a peduncle or rarely bearing a flower:
    11 Seeds globose with irregular, raised, pale ridges, minutely punctate in between; lamina auriculate at junction with sheath . . . . 10. *C. zambesica*
    11 Seeds long-ellipsoid, with deep transverse grooves and strong ridges converging towards embryotega; lamina attenuate at junction of sheath . . . . 11. *C. petersii*
6 Capsule triangular with 3 one-seeded, subglobose locules, or with only the 2 ventral locules bearing seeds, the dorsal locule smaller or absent, or indehiscent and tuberculate (in *C. erecta*):
  12 Cymes 2 in each spathe, both lower and upper well developed; leaf-margin white, undulate; spathes sessile, clustered at the apices; flowers large; Springbok Flats, Transvaal . . . . 16. *C. bella*
  12 Cymes 1 in each spathe, the upper aborted:
    13 Capsule with all 3 locules (occasionally 2 when dorsal locule is aborted) dehiscent, smooth:
      14 Annuals; spathes sessile, congested above, falcate-rostrate; petals pale orange; capsule oblong-globose; annual weeds . . . . 12. *C. aspera*
      14 Perennials (chamaephytes):
        15 Spathes solitary, rarely with a younger one above it; leaves flat, narrowly linear, c. 4 mm broad, tapered into the sheath; plants tenuous, saxicolous . . . . 15. *C. modesta*
        15 Spathes clustered and subsessile at the apices of the stems; leaves folded, linear to oblong, up to c. 10 mm broad, eared at the base, abruptly narrowed into the sheath . . . . 13. *C. livingstonii*
    13 Capsule with the dorsal locule indehiscent, tuberculate . . . . 14. *C. erecta*

    1. **Commelina subulata** *Roth*, Nov. Pl. Sp. 23 (1821); C. B. Cl. in A. DC., Monogr. Phan. 3: 148 (1881), and in F.C. 7: 9 (1897), and in F.T.A. 8: 38 (1901); Morton in J. Linn. Soc., Bot. 60: 189 (1967); Brenan in F.W.T.A. edn 2, 3: 47 (1968); Schreiber et al. in F.S.W.A. 157: 9 (1969); Hilliard & Burtt in Notes R. bot. Gdn Edinb. 37: 294 (1979). Type: India, *Heyne* (B, holo.; K, iso., PRE, photo.!).

C. violacea C.B. Cl. in F.T.A. 8: 39 (1901). Syntypes: South West Africa/Namibia, in marshy places at Olukonda, *Schinz* 21, 33 (Z; K!; BM; US); cf. Norlindh in Bot. Not. 1948: 17 (1948).

Short-lived annuals flowering precociously, at first with a simple, erect stem but becoming many-stemmed and dome-shaped, up to 0,25 m tall, with long internodes, glabrous except for ciliate open leaf-base and margin of spathe. *Leaves* linear, 40–70 × 3 mm, margins smooth, apex acute. *Spathe* folded, falcate, sharply recurved, c. 5–10 mm long, subsessile, often striate. *Cyme* solitary, 2–3-flowered, flowers in turn just exserted at anthesis, small; petals yellow, orange, pink, apricot, blue or brownish violet. *Stamens* 3, fertile, coloured like petals; staminodes yellow. *Ovary* with 2 ovules in each ventral cell, dorsal cell with 1 or 0 ovules. *Capsule* ovoid, acute, often beaked, with dorsal locule shorter than ventral ones and with a solitary larger seed; seeds of ventral locule subrectangular, with 3 cross-furrows, scarcely tuberculate, smooth below.

Widespread in tropical Africa to the Arabian Peninsula and southern India; reaching the northern parts of Southern Africa. Weed-like in behaviour, short-lived, in seasonally wet areas such as margins of pans. Map 20.

Vouchers: *De Winter & Marais* 4925; *Hilliard & Burtt* 9924; *Leistner* 3016; *Schlechter* 4249; *Smith* 839; *Wild & Drummond* 7039; *Volk* 11930.

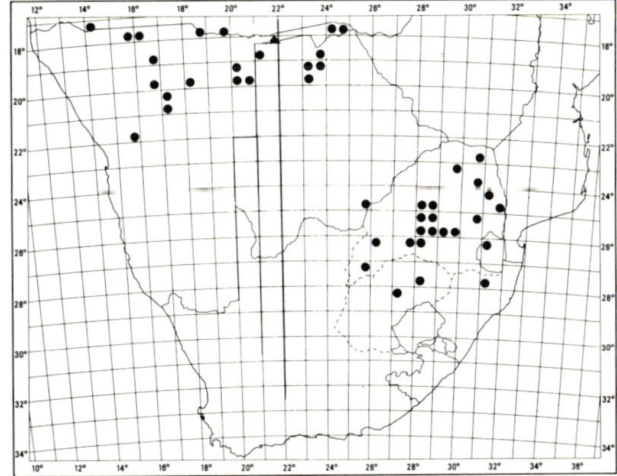

MAP 20.— ● **Commelina subulata**
       ▲ **Commelina fluviatilis**

2. Commelina fluviatilis *Brenan* in Mitt. bot. StSamml. Münch. 6: 253 (1967); Schreiber et al. in F.S.W.A. 157: 8 (1969). Type: Zambia, Mpika district, Luitikila River, *Richards* 14984 (K, holo.!).

Aquatic or semi-aquatic, glabrous herb with long, floating or trailing stems rooting at nodes, side-branches short, erect. *Leaves* few, emergent, narrowly linear, 50–250 × 1–3,5 mm, sheath up to 30 mm long, striate. *Spathes* usually solitary, terminal, pedunculate, folded, 10–20 mm long, straight or falcate, acute to acuminate, striate. *Cymes* with upper exserted, 1-flowered, lower c. 4-flowered, shortly exserted at anthesis. *Flowers* small with petals long-clawed, c. 15 mm long (in dried flowers), pale mauve, blue or white. *Stamens* with 3 lower fertile; staminodes 3, shorter. *Ovary* 5-ovuled; style c. 12 mm long. *Capsule* oblong-apiculate, up to 7 mm long 2(–4?)-seeded; seeds greyish brown, usually ellipsoid, 2–3 mm long, irregularly foveolate.

Recorded from Zambia, Botswana and northern South West Africa/Namibia; on riverbanks trailing into the water or in moist sandy places. Map 20.

Vouchers: *De Winter & Marais* 4884; *Dinter* 7201; *Gibbs Russell* 2820.

3. Commelina diffusa *Burm. f.*, Fl. Ind. 18, t. 7, f. 2 (1768); Merrill in J. Arn. Arb. 18: 64 (1973); Morton in J. Linn. Soc., Bot. 55: 521, f. 18 (1956); l.c. 60: 181, f. 3 (1967); Brenan in F.W.T.A. edn 2, 3: 47, f. 332 (1968). Type: India, Burman coll. (G).

Annual or perennial hygrophilous spreading herbs sometimes covering large stretches of ground, creeping stems forming short, erect side branches, rooting at nodes. *Leaves* linear to ovate or oblong, 50–140 × 15–25 mm, apex acuminate, glabrous or puberulous, margins often encrusted with white pustules. *Spathes* pedunculate, folded, broad and acute to long and attenuate. Upper cyme exserted on an erect, glabrous or pubescent peduncle, (1–)2(–3)-flowered, rarely aborted; lower peduncle pubescent, with 3–8 bisexual flowers. *Sepals* white, membranous. *Petals* blue or white. *Stamens* with yellow, red, blue or black filaments and anthers. *Capsule* quadrate; 2 ventral locules 2-seeded, dorsal locule indehiscent, fusiform, 1-seeded, or empty; seeds reniform, reticulate with tuberculate hexagonal areoles.

A widely distributed pantropical innocuous hygrophytic weed; in Southern Africa recorded from South West Africa/Namibia, Botswana, Transvaal, Swaziland, Natal to eastern and southern Cape.

Two subspecies are recognized:

1 Leaves ovate, c. 50 × 15 mm; usually soft, mesophytic plants; widespread . . . . (a) subsp. *diffusa*
1 Leaves linear, c. 100–150 × 10 mm; more robust plants, found mainly in the northern parts of Southern Africa and somewhat further north . . . . (b) subsp. *scandens*

(a) subsp. **diffusa.**

C. nudiflora L., Sp. Pl. 41 (1753) partly, excl. lectotype of Merrill, l.c.; C. B. Cl. in F.C. 7: 8 (1897), and in F.T.A. 8: 36 (1902).

C. werneana Hassk. in Schweinf., Beitr. Fl. Aethiop. 206, 295 (1867). *C. nudiflora* var. *werneana* (Hassk.) C. B. Cl. in F.C. 7: 9 (1897). Type: ? Egypt, *Werne* (B).

The typical variety is a smaller and weak-stemmed, more mesophytic plant with soft, ovate leaves c. 50 mm long and shorter nodes.

Pantropical; recorded from the northern, eastern and southern parts of Southern Africa. Fig. 5. Map 21.

Vouchers: *Boucher* 3474; *Scheepers* 4; *Smith* 2541; *Strey* 8496; *Ward* 5439, 6254.

(b) subsp. **scandens** *(C.B. Cl.) Oberm.* in Bothalia 13: 437 (1981).

C. scandens Welw. ex C. B. Cl. in A. DC., Monogr. Phan. 3: 146 (1881), and in F.T.A. 8: 37 (1901); Schreiber et al. in F.S.W.A. 157: 9 (1969). Type: Angola, Pungo Andongo, banks of River Cuanza near Nbilla, *Welwitsch* 6642 (BM, holo.).

More robust, long-stemmed herb with linear to narrowly oblong, attenuate leaves up to c. 0,14 m long.

Common on the banks of the Okavango River and swamps amongst dense vegetation in northern South West

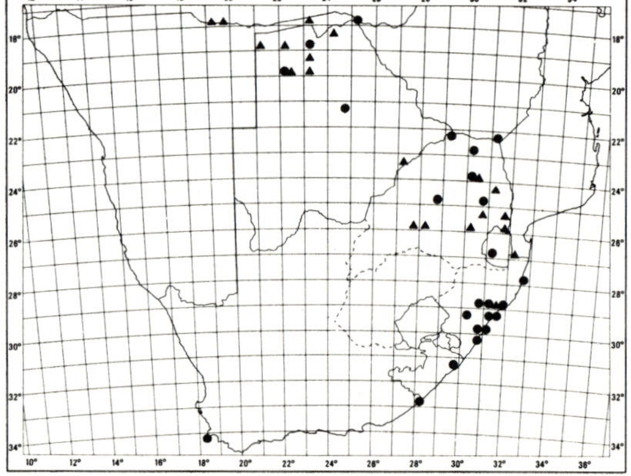

MAP 21.— ● **Commelina diffusa** subsp. **diffusa**
    ▲ **Commelina diffusa** subsp. **scandens**

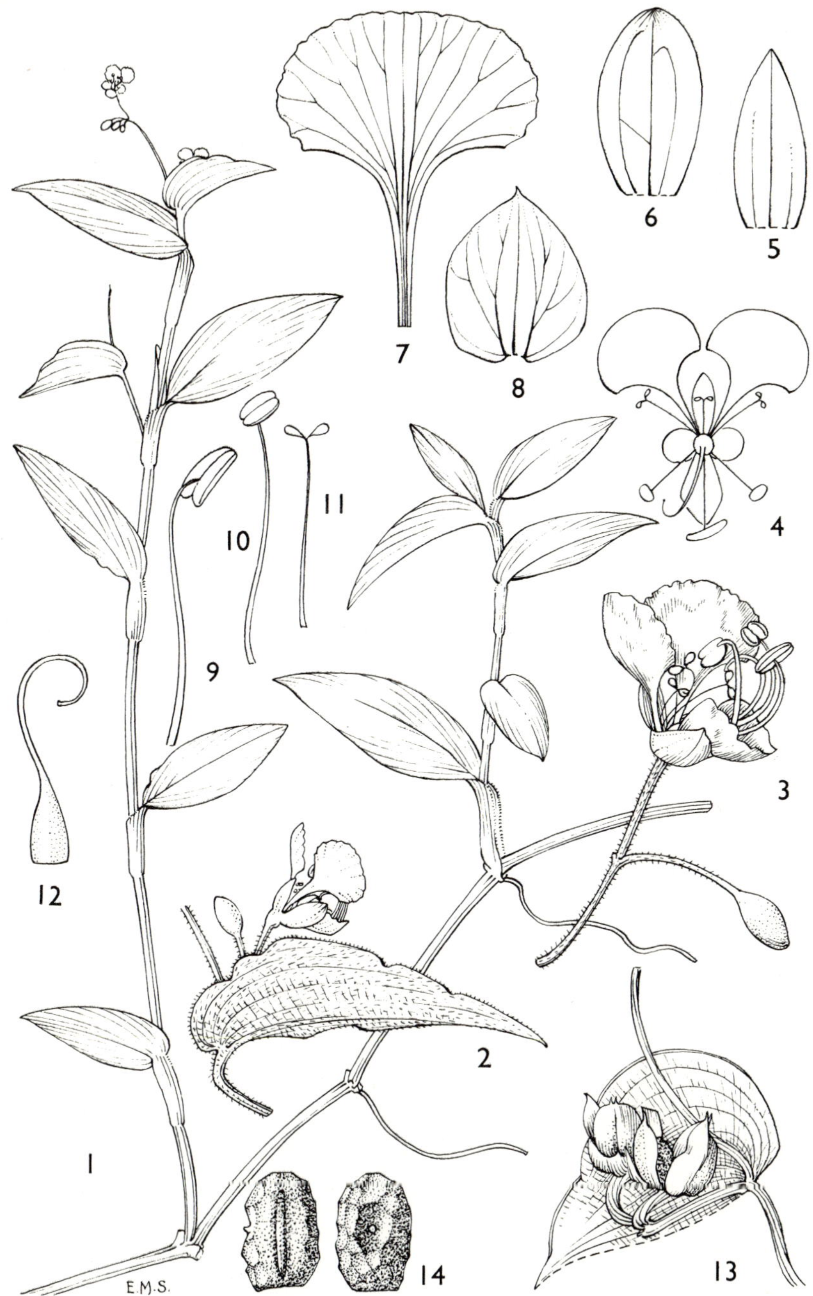

FIG. 5.—**Commelina diffusa**: 1, habit, × 0,6; 2, inflorescence, × 2; 3, flower and bud, × 6; 4, floral diagram; 5, dorsal sepal, × 8; 6, sepal, × 8; 7, petal, × 8; 8, keeled petal, × 8; 9, stamen, × 8; 10, stamen (one of pair), × 8; 11, staminode, × 8; 12 gynoecium, × 8; 13, fruit, × 2; 14, seeds, × 6; (1 from *Gillett* 15348; 2–12, from *Gillett* 15295a; 13, 14 from *Okeke* FHI 18228. Reproduced from the 'Flora of West Tropical Africa' with permission of the Director of the Royal Botanic Gardens, Kew.

Africa/Namibia and Botswana; also recorded from Transvaal and Natal; further north in Angola to western tropical Africa. Map 21.

Vouchers: *Codd* 5507; *Marais & De Winter* 4524; *Pooley* 1360; *Scheepers* 1204; *Smith* 1332, 1543; *Wells* 2479; *Wild & Drummond* 7107.

4. **Commelina africana** *L.*, Sp. Pl. 41 (1753); Red., Liliac. 4, t. 207 (1808); Ker-Gawl. in Curtis's bot. Mag. 35, t. 1431 (1811); C. B. Cl. in A. DC., Monogr. Phan. 3: 164 (1881), and in F.C. 7: 9 (1897), and in F.T.A. 8: 45 (1901); Marloth, Fl. S. Afr. 4: 68, t. 18 (1915); Morton in J. Linn. Soc., Bot. 515 (1956), and in l.c. 60: 147 (1967); Adamson in Adamson & Salter, Fl. Cape Penins. 159 (1950); Brenan in Mitt. bot. StSamml. Münch. 5: 203 (1964), and in F.W.T.A. edn 2,3: 45 (1968); Schreiber et al. in F.S.W.A. 157: 6 (1969). Type: a cultivated plant in hort. Upsal. (LINN 65.3, holo., PRE, photo.!).

Spreading perennial herbs (chamaephytes) up to 0,5 m tall, glabrous or variously pubescent; rootstock hard, woody, with hard, thick, long roots. *Leaves* variable, oblong to linear, flat or folded, up to 120 mm long but usually smaller, glabrous to variously hairy. *Spathes* folded, dry inside, solitary, pedunculate; apex acute to long-acuminate, often falcate. Both cymes well developed. *Flowers* varying in size, petals yellow. *Ovary* 5-ovuled. *Capsule* oblong, with dorsal locule indehiscent, 1-seeded or often empty, ventral locules dehiscent, each 2-ovuled but usually 1-seeded, lower ovules aborting. *Seeds* oblong-ellipsoid, reticulate.

Widespread in Africa, Madagascar and the Arabian Peninsula; in forests, savanna, grassland, etc., often on rocky outcrops. Very common in Southern Africa.

This very variable species was divided into twelve varieties by Brenan in Mitt. bot. StSamml. Münch. 5: 199–222 (1964), of which nine have been recorded for Southern Africa. As several of these varieties are not clearly distinguished in our region it was decided to divide the Southern African material into 4 varieties:

1 Leaves flat, oblong, narrowed at the base, usually glabrous .... (a) var. *africana*
1 Leaves folded, usually linear, not or hardly narrowed at the base:
  2 Spathes usually narrow, long-acuminate; leaves glabrous or glabrescent, broadly to narrowly linear:
    3 Flowers small (upper petals c. 8 mm long); leaves usually more than 50 mm long .... (b) var. *lancispatha*
    3 Flowers large (upper petals c. 15 mm long); leaves narrow, short, falcate, the margins often crenulate .... (c) var. *barberae*
  2 Spathes usually short and broad; leaves pubescent .... (d) var. *krebsiana*

(a) var. **africana.**

C. welwitschii C.B. Cl. in A. DC., Monogr. Phan. 3: 175 (1881). Type: Angola, Humpata, *Welwitsch* 6586 (K, BM).

Mesophytic, usually glabrous with flat oblong leaves.

From the Cape Peninsula eastwards along the eastern Cape to Natal and the eastern Transvaal and further north to tropical Africa; in undergrowth along forest edges and riverbanks. Map 22.

Vouchers: *Arbuthnot* sub PRE 37948; *Codd* 8404; *Compton* 31792; *Pegler* 2191; *Strey* 4510.

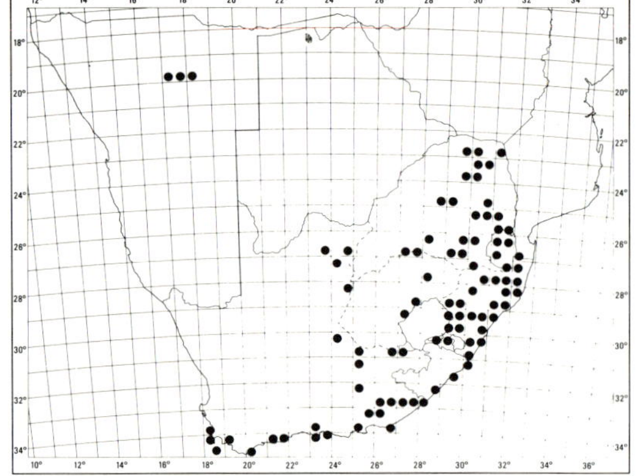

MAP 22.—**Commelina africana** var. **africana**

(b) var. **lancispatha** *C.B. Cl.* in F.C. 7: 10 (1897); Phillips in Flower. Pl. Afr. 9, t. 321 (1929); Brenan in bot. StSamml. Münch. 5: 211 (1964). Type: E. Cape, Zuurberg Range, *Drège* 8779 (K!).

C. dinteri Mildbr. in Notizbl. bot. Gart. Mus. Berl. 9: 253 (1925). Type: South West Africa/Namibia, Gaub, *Dinter* 2445 (B).

C. krebsiana var. *glabriuscula* T. Norl. in Bot. Notiser 1948: 20 (1948). *C. africana* var. *glabriuscula* (T. Norl.) Brenan in Mitt. bot. StSamml. Münch. 5: 217 (1964). Type: Zimbabwe, near Inyanga, *Fries, Norlindh & Weimarck* 3141 (PRE, iso.!).

Leaves linear, folded, often falcate, apex attenuate, not, or only somewhat narrowed at the base, glabrous or glabrescent. *Spathe* usually narrow, attenuate towards the apex, often falcate. Fig. 6: 2.

Widespread and common in many areas of Southern Africa, including the N.W. Cape, western Orange Free State and western Transvaal. Map 23.

Vouchers: *Hanekom* 2241; *Marais* 172; *Rogers* 2191; *Smith* 134; *Van Nouhuys* sub PRE 7938; *Werger* 236.

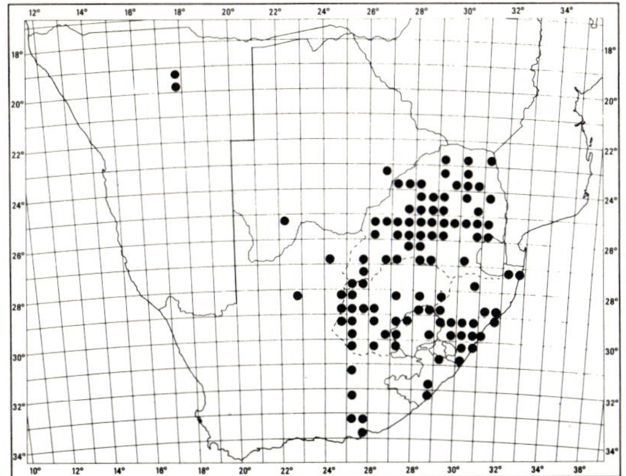

MAP 23.— **Commelina africana** var. **lancispatha**

(c) var. **barberae** *(C.B. Cl.) C.B. Cl.* in F.C. 7: 10 (1897); Brenan in Mitt. bot. StSamml. Münch. 5: 215 (1964). Type: Vaal River, *Barber* 10/75 (K, holo.).

C. karooica C.B. Cl. var. *barberae* C.B. Cl. in A. DC., Monogr. Phan. 3: 166 (1881).

C. karooica C.B. Cl. in A. DC., Monogr. Phan. 3: 166 (1881), and in F.C. 7: 10 (1897). Type: Cape, Griqualand West, between Griqua Town and Witte Water, *Burchell* 1999 (K, syn.).

This dainty variety has linear, short, often crisped, falcate leaves. The flowers are relatively large: the upper petals may reach a length of c. 15 mm. As in var. *lancispatha* the spathe is attenuate.

Var. *barberae* has been recorded mainly from drier regions. Map 24.

Vouchers: *Bolus* 10823; *Leistner* 1260; *Reyneke* 437; *Rodin* 3520.

(d) var. **krebsiana** *(Kunth) C.B. Cl.* in A. DC., Monogr. Phan. 3: 164 (1881); Brenan in Mitt. bot. StSamml. Münch. 5: 216 (1964). Type: Cape, *Krebs* (B, holo.).

C. krebsiana Kunth, Enum. Pl. 4: 40 (1843); C. B. Cl. in F.C. 7: 10 (1897).

C. barbata Lam. var. *villosior* C.B. Cl. in A. DC., Monogr. Phan. 3: 167 (1881).

C. africana var. *villosior* (C.B. Cl.) Brenan in Mitt. bot. StSamml Münch. 5: 207 (1964). Lectotype: Natal, 30 °S., 1855, *Sutherland* s.n. (K!).

C. boehmiana K. Schum. in Engl., Pflanzenw. Ost-Afr. (C): 135 (1895). *C. africana* var. *boehmiana* (K. Schum.) Brenan in Mitt. bot. StSamml Münch. 5: 213 (1964). Type: Tanzania, Gonda, *Boehm* 12 (B).

C. africana var. *brevipila* Brenan in Mitt. bot. StSamml. Münch. 5: 219 (1964). Type: Zambia, Kapiri Mposhi, *Fanshawe* 1823 (K, holo.!).

C. africana var. *milleri* Brenan in Mitt. bot. StSamml. Münch. 5: 221 (1964). Type: Zimbabwe, Besna Kobila, *Miller* 4061 (K, holo.!).

The pubescent plants, which vary from hispid to densely tomentose, are here placed in this variety. Scabrid, long, several-celled setae may also be present. The leaves are often broader and shorter than in var. *lancispatha*.

This variety is widely distributed. Map 25.

Vouchers: *Adamson* D255; *Bolus* 6417; *Burtt Davy* 1274; *Giess & Müller* 14819; *Hutchinson* 2658; *Mauve* 5038.

5. **Commelina rogersii** *Burtt Davy* in Jl S. Afr. Bot. 4: 125 (1938). Type: Transvaal: Pietersburg, *F. A. Rogers* 14142 (K, holo.!; PRE, iso.!).

Erect or spreading, wiry, puberulous chamaephytes up to c. 0,3 m tall. *Stems* with

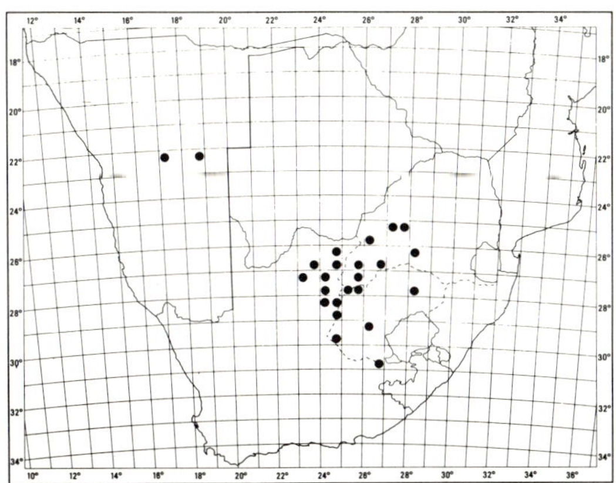

MAP 24.— **Commelina africana** var. **barberae**

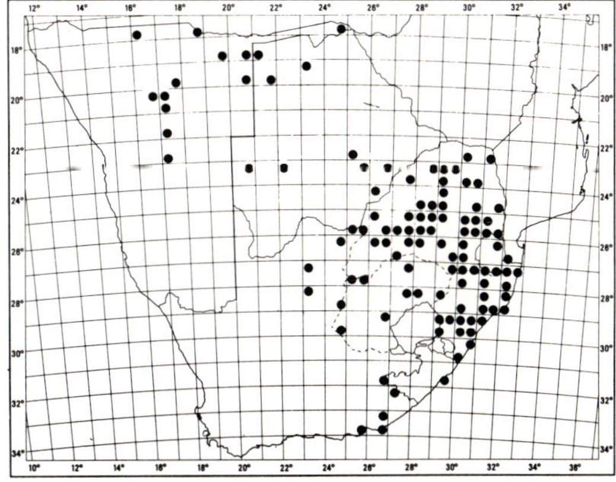

MAP 25.— **Commelina africana** var. **krebsiana**

elongated thin, woody internodes below. *Leaves* imbricate on short side shoots above, narrowly linear, up to c. 30 × 3 mm, striate, margins raised, white with few to many long white septate setae above mouth of short sheath. *Spathes* on patent peduncles up to 15 mm long, folded, narrowly ovate-acuminate, falcate, striate, setose, margins raised, lined on the inside with white flat hairs. *Upper cyme* on a patent stipe 10–15 mm long, occasionally fertile. *Lower cyme* c. 4-flowered. *Sepals* membranous, speckled with brown. *Petals* yellow. *Capsule* narrowly ovoid, c. 6 mm; seeds absent.

A rarely collected species recorded from grasslands around Pietersburg and with one collection from the Soutpansberg near Louis Trichardt; flowering January–April. Map 26.

Vouchers: *Bredenkamp & van Vuuren* 114; *Moss* 15689; *Rodin* 4001; *Schlechter* 4350.

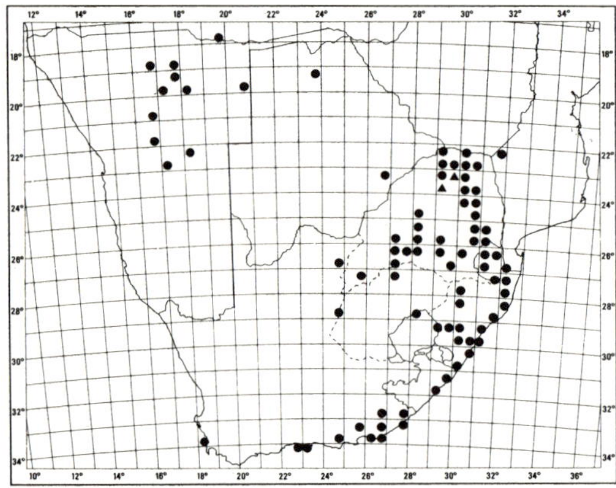

MAP 26.— ● **Commelina benghalensis**
              ▲ **Commelina rogersii**

6. **Commelina benghalensis** *L.*, Sp. Pl. 41 (1753); C. B. Cl. in A. DC., Monogr. Phan. 3: 159 (1881), and in F.C. 7: 9 (1897), and in F.T.A. 8: 41 (1901); Phillips in Flower. Pl. S. Afr. 2, t. 42 (1922); Morton in J. Linn. Soc., Bot. 55: 519, f. 15 (1956); l.c. 60: 176 (1967); Henderson & Anderson in Mem. bot. Surv. S. Afr. 37, t. 31 (1966); Brenan in F.W.T.A. edn 2, 3: 48 (1968); Schreiber et al. in F.S.W.A. 157: 7 (1969). Types: India (LINN. 65.4, 65.5, PRE, photos!).

Spreading annual herbs, sparsely and shortly pubescent, sometimes bearing subterranean runners with cleistogamous, reduced flowers; roots thin. *Leaves* ovate, up to c. 80 × 30 mm, pale apple green, apex obtuse to acute, base abruptly narrowed into a sheath-

ing petiole beset with long, red, or rarely colourless, several-celled setae at the mouth. *Spathes* subsessile, clustered at apices of branches, obliquely fused, triangular, 10 × 15 mm, apex short, acute. *Cymes* 2, the upper one often suppressed later in the season. *Flowers* small, petals a deep ink-blue. *Capsule* 5-seeded; dorsal locule 1-seeded, indehiscent, shed with part of adjoining wall of the 2-seeded ventral locules; seeds oblong, 4 mm long in dorsal locule, 2 mm long in ventral locules, dark brown with deep cross furrows. Fig. 6: 3.

A common, troublesome, widespread weed found throughout Southern Africa, tropical Africa and Asia; naturalized in North and South America. Because of the subterranean, seed-bearing capsules it is difficult to eradicate in cultivated lands. Map 26.

Vouchers: *Compton* 27457; *De Winter & Marais* 4489; *Pont* 364; *Ross* 2227; *Scheepers* 57.

7. **Commelina forskaolii** *Vahl*, Enum. Pl. 2: 172 (1806); C. B. Cl. in A. DC., Monogr. Phan. 3: 168 (1881), and in F.T.A. 8: 44 (1901); Morton in J. Linn. Soc., Bot. 55: 522, f. 19 (1956), l.c. 60: 185 (1967); Maheshwari & Baldev in Phytomorphology 8: 277 (1958); Brenan in F.W.T.A. edn 2, 3: 48 (1968); Schreiber et al. in F.S.W.A. 157: 8 (1969). Type: Arabia, *Forsskål* (C, holo.).

Spreading annual herb or chamaephyte with long, trailing stems rooting at nodes and forming short erect branches and often subterranean stolons bearing reduced cleistogamous flowers. *Leaves* folded, linear to oblong, 20–70 mm long, margins wavy, pale greyish green, glabrous or puberulous. *Spathes* 1–2, terminal and leaf-opposed, shortly pedunculate, obliquely funnel-shaped with a short, acute apex, c. 10 mm long. Both cymes developed. *Petals* blue. *Lateral stamens* with filaments more or less winged. *Capsule* sub-obovoid; dorsal locule indehiscent, striate-muricate, 1-seeded; 2 ventral locules 2-ovulate but usually forming only 1 seed; testa smooth.

Widespread in Africa, Arabia, Socotra, Madagascar and India; recorded from the northern parts of Southern Africa in sandy, dry bushveld vegetation. Map 27.

Vouchers: *De Winter & Wiss* 4324; *Pole Evans* 4561; *Ross* 2343; *Van Son* sub TRV 29040; *Wild* 5093; *Zwanziger* 751.

Maheshwari & Baldev in Phytomorphology 8: 284–285 (1958) observed an inverse relationship in the development of aerial and subterranean branches. Under drought conditions the plants developed poorly above ground but the underground branches showed luxuriant

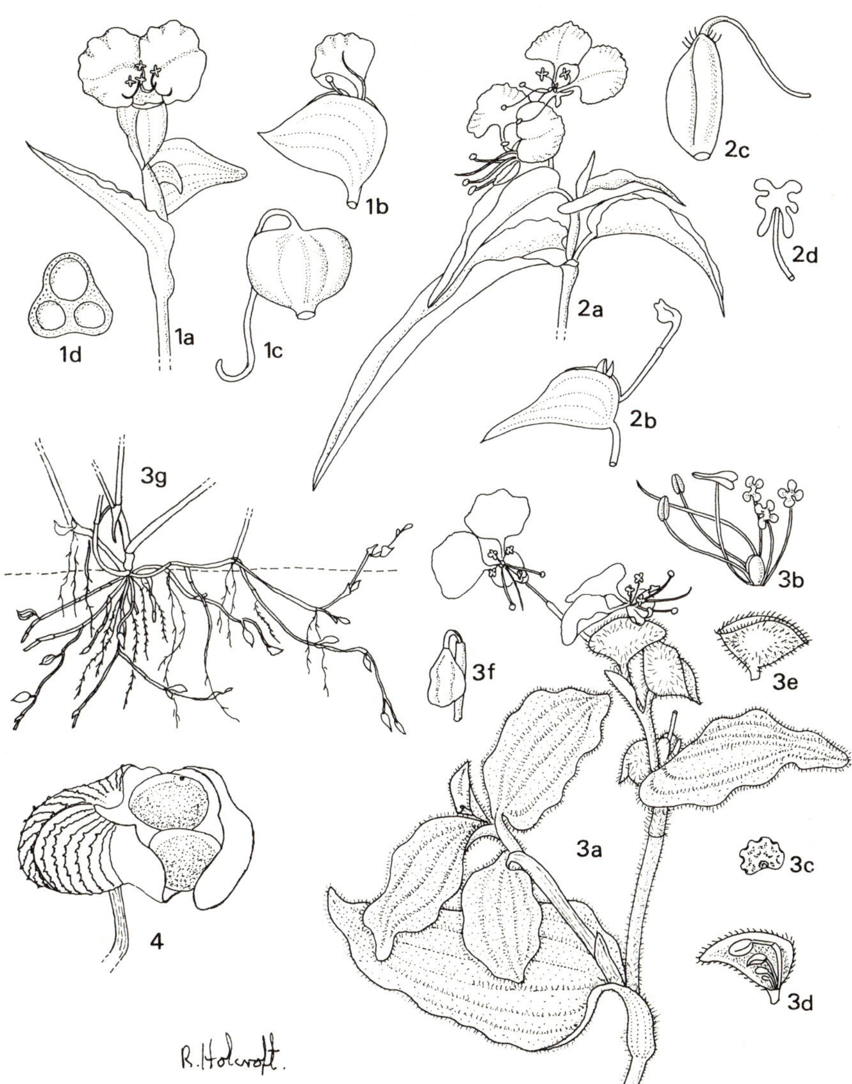

FIG. 6.—1, **Commelina livingstonii**: 1a, flower exserted from spathe, (lower cyme only developed), front view, × 0,8; 1b, side view, × 0,8; 1c, capsule with style, × 1,5; 1d, transverse section through capsule, × 1 (after C. Letty in Flower Pl. Afr. 9, t. 323); 2, **C. africana** var. **lancispatha**, 2a, flowering branch with both cymes developed, × 0,8; 2b, spathe with buds, × 0,8; 2c, immature capsule, × 2; 2d, staminode, × 3 (after C. Letty in Flower. Pl. Afr. 9. t. 321); 3, **C. benghalensis**, 3a, flowering branch, with both cymes developed, × 0,8; 3b, stamens and pistil, × 2; 3c, seed, × 1,5; 3d, young inflorescence with half of spathe removed, × 0,8; 3e spathe, × 0,8; 3f, capsule, × 3; 3g, root system showing underground stolons bearing reduced, cleistogamous flowers and fertile seeds, × 0,3 (figs 3a–f, after M. Page in Flower. Pl. Afr. 2, t. 42); 4, **C. erecta**, capsule bursting, showing indehiscent dorsal locule and dehiscent ventral locules with freed seeds, × 4,5 (*Moll* 2907).

growth and seed development. The opposite occurred in well-watered plants. This suggests "an adaptation to survival under unfavourable conditions". The seed from the underground flowers proved to be more fertile than that of the aerial flowers, which may be self-pollinated. The authors neither mention nor depict the warty surface of the indeshiscent dorsal locule, which was observed in material from Africa as well as the Arabian Peninsula.

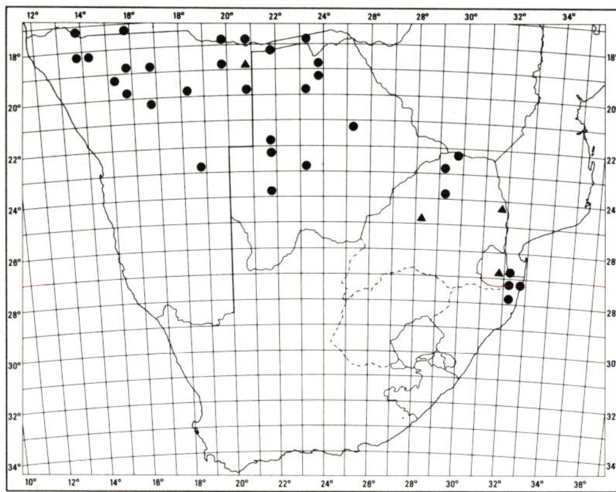

MAP 27.— ● **Commelina forskaolii**
        ▲ **Commelina imberbis**

8.   **Commelina imberbis** *Hassk.* in Schweinf., Beitr. Fl. Aethiop. 209, 295 (1867); C. B. Cl. in F.T.A. 8: 49 (1901); Brenan in F.W.T.A. edn 2, 3: 48 (1968). Type: Ethiopia, Togodele, *Ehrenberg* (B†).

Weedy annuals with long, straggling stems rooting at the nodes, usually glabrous. *Leaves* broadly linear-acuminate, cordate and amplexicaul at the base, 50–90 × 10–15 mm, pale green, margins often crenulate, minutely and sparsely setaceous with the setae appressed to the leaf-surface. *Spathes* pedunculate, short and broad, free or fused basally, acute, c. 20 × 10 mm. *Upper cyme* 0 or represented by a peduncle. *Flowers* with blue petals. *Capsule* quadrate with the dorsal locule aborted or rarely 1-seeded; seeds cylindric-ellipsoid, 3 mm long, smooth, farinose with a short thick hilum, slightly marbled.

South West Africa/Namibia, Transvaal and Swaziland; widespread in tropical Africa, from Ethiopia and tropical West Africa; also recorded from the Arabian Peninsula. Weed-like in behaviour. Said to be weed-killer resistant. Map 27.

Vouchers: *Brenan & Vahrmeijer* 14259; *Gertenbach* 5416; *Wild & Drummond* 7021.

The relationship between this species and *C. kotschyi* Hassk., described at the same time, needs further investigation.

9.   **Commelina eckloniana** *Kunth*, Enum. 4: 57 (1843); C. B. Cl. in A. DC., Monogr. Phan. 3: 174 (1881), and in F.C. 7: 11 (1897); Phillips in Flower. Pl. S. Afr. 9, t. 326 (1929). Type: Cape, *Ecklon* (K, iso., PRE photo.!).

C. weimarckiana Norl. in Bot. Notiser 1849: 22 (1948). Type: Zimbabwe, Inyanga, *Norlindh & Weimarck* 4269 (S, holo.; PRE, iso.!).

Chamaephytes, glabrous except for some long white straggling hairs; stems annual spreading, arising from a knobbly rootstock bearing long, fusiform, dark roots covered by root hairs. *Stems* c. 200–350 mm long, internodes up to c. 90 mm long. *Leaves* linear-acuminate to ovate-acuminate, 60–100 mm long, (–3)8–25 mm broad, pale grey-green. *Spathes* solitary, obliquely funnelform, broad and short, recurved, 10–15 mm long, on straight, patent peduncles 10–25 mm long. *Upper cyme* exserted, 1-flowered, lower with flowers just exserted. *Petals* blue. *Capsule* quadrate, flat, c. 5 mm long; the dorsal locule not developed, the raised margin constricted between 4 seeds; seeds globose, tuberculate, c. 2 mm diam.

Widespread in Southern Africa; found as far north as Ethiopia; usually growing in rocky habitats. Map 28.

Vouchers: *Galpin* 1188; *Scharf* 1155; *Strey* 4979; *Thode* A468; *Van der Schijff* 1526.

Sometimes eaten as spinach and known in Venda as damba.

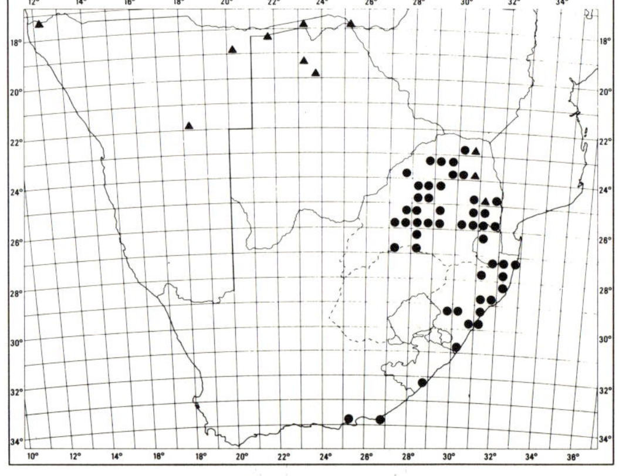

MAP 28.— ● **Commelina eckloniana**
        ▲ **Commelina zambesica**

10.   **Commelina zambesica** *C.B. Cl.* in A. DC., Monogr. Phan. 3: 161 (1881), and in F.T.A. 8: 43 (1902); Brenan in Kew Bull. 15: 209 (1961); Morton in J. Linn. Soc., Bot.

60: 190 (1967); Brenan in F.W.T.A., edn 2,3: 48 (1968). Syntypes: Zimbabwe, near Zambesi R., *Kirk* s.n. (K!); near Lake Nyassa, *Simons* (BM).

Straggling, robust, pale green annual herbs. *Leaves* narrowly elliptic, attenuate above and below, c. 100 × 25 mm, scabrid, especially so along margins, sparsely hairy (hairs multicellular), sheath up to 20 mm long, auriculate at the apex, ciliate. *Spathe* on a firm erect peduncle c. 10 mm long, funnel-form, broadly ovate, apex abruptly acute, hispid. *Upper cyme* well-developed or reduced to a short stalk; lower cyme c. 3-flowered. *Flowers* with petals, stamens and staminodes ink-blue; ovary smooth, green, 3-celled, 5-ovuled. *Capsule* asymmetrically obovoid, obtuse; dorsal locule with 1 large seed; 2 ventral locules with smaller seeds; seeds globose to oblong-globose, brown with some irregular, raised, white ridges converging towards embryotega and minutely papillate.

Widespread in West and East tropical Africa to Zimbabwe, South West Africa/Namibia and northern Transvaal. Map 28.

Vouchers: *Codd* 6827; *Edwards* 4539; *Smith* 2355.

In appearance and size resembling *C. erecta* L. (no. 14), but the capsule is 5-seeded (not 3-seeded) and the seeds are brownish with whitish ribs converging towards the embryotega, whereas in *C. erecta* they are smooth and farinose.

11. **Commelina petersii** *Hassk.* in Peters, Reise Mossamb., Bot. 522 (1864); C. B. Cl. in A. DC., Monogr. Phan. 3: 169 (1881), and in F.T.A. 8: 50 (1901); Brenan in F.W.T.A. edn 2, 3: 48 (1968). Type: Mozambique, *Peters* (B†).

Perennial herb, erect or scrambling, up to 0,8 m tall with long internodes; roots hard, covered by roothairs. *Leaves* narrowly ovate-acuminate, up to 100 × 20 mm, abruptly narrowed at base into a pseudo-petiole and a long membranous sheath, dark green above, paler below, laxly and minutely asperous. *Spathe* solitary, falcate, c. 25–30 mm long, acuminate, shortly fused at base, mucilaginous inside; on a peduncle 20–30 mm long. *Upper cyme* on a long exserted peduncle, bearing a male flower or barren. *Flowers* with petals blue, sometimes with a mauve tinge. *Capsule* oblong; ventral locules 2-seeded; seeds long-ellipsoid with deep transverse grooves and minutely tuberculate ridges.

Recorded from Ethiopia and West Tropical Africa to the northern part of the Flora area, where it is rare. Map 29.

Vouchers: *Merxmüller & Giess* 30579; *Rogers* 20986; *Schinz* 6 (K).

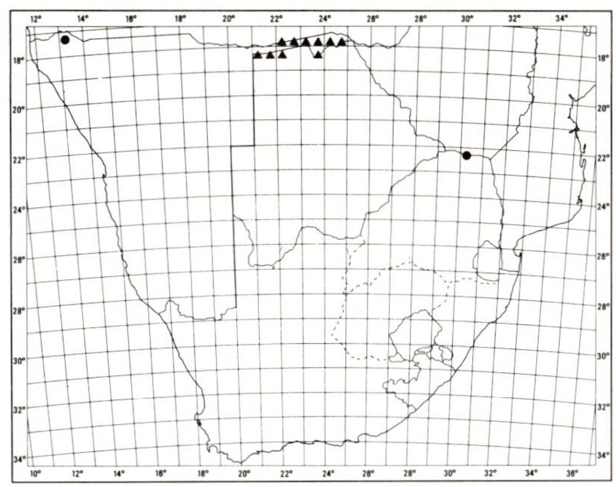

MAP 29.— ● **Commelina petersii**
     ▲ **Commelina aspera**

12. **Commelina aspera** *Benth.* in Fl. Nigrit. 542 (1849); C. B. Cl. in A. DC., Monogr. Phan. 3: 180 (1881); Dur. & Schinz, Consp. Fl. Afr. 5: 423 (1895); C. B. Cl. in F.T.A. 8: 56 (1901); Morton in J. Linn. Soc., Bot. 55: 518, f. 14 (1956); l.c. 60: 176 (1967); Brenan in F.W.T.A. edn 2, 3: 50 (1968); Schreiber et al. in F.S.W.A. 157: 7 (1969). Type: Upper Guinea, Gold Coast, Accra, *Don* (BM).

Annuals, flowering precociously, simple-stemmed at first, becoming bushy and straggling with age, densely asperous (the setae uni- to multi-cellular). *Leaves* linear-acuminate, c. 70 mm long, folded. *Spathes* clustered apically, subsessile, funnel-form, obliquely triangular, c. 12 mm long, apex sharply recurved. *Flowers* with petals yellow, apricot or white, only just exserted at anthesis. *Capsule* small, ovoid, apiculate, trilocular, locules papery, one-seeded, separating and splitting; seeds ellipsoid, 3 mm long, hilum forming a long, white double ridge; testa with sparse short white tubercles irregularly spaced.

Botswana and northern South West Africa/Namibia; also northwards to Zimbabwe and other parts of tropical Africa; in savanna and grassland, usually in sandy soil, often a weed in cultivated lands and open spaces. Map 29.

Voucher: *Kruger* s.n. sub PRE 57930.

13. Commelina livingstonii C.B. Cl. in A. DC., Monogr. Phan. 3: 190 (1881), incl. var. b, and in F.C. 7: 11 (1897), and in F.T.A. 8: 59 (1901); Schreiber et al. in F.S.W.A. 157: 9 (1967). Type: Zimbabwe, near Zambesi River, *Kirk* (K, holo., PRE, photo.!).

C. erecta L. subsp. *livingstonii* (C.B. Cl.) Morton in J. Linn. Soc., Bot. 60: 184 (1967).

C. livingstonii var. *villosa* C.B. Cl. in A. DC., Monogr. Phan. 3: 190 (1881). Syntypes: N.W. Cape, near Kuruman, Hamapery, *Burchell* 2515; Apies River, *Burke* s.n.; Mooi River, *Burke* 339; without locality, *Zeyher* 1729 (all K, PRE, photos!).

C. albescens sensu C.B. Cl. in F.C. 7: 11 for *Galpin* 599; sensu Phillips in Flower. Pl. Afr. 9, t. 323 (1929); non Hassk.

Chamaephytes with spreading branched stems c. 100–300 mm tall, from a woody, hard crown; roots thick, hard and long. *Stems* angled, shortly setose. *Leaves* narrowly ovate-acuminate, c. 70 × 20 mm, auriculate at junction with sheath, usually folded, greyish green, glabrescent to pubescent with long and short setae. In some vernal stages the leaves may be long and narrow, folded. *Spathes* fused, (1)–3, clustered apically, subsessile, broad and short, apiculate, c. 15 mm long, finely ribbed, setulose. Only the lower cyme developed, 3–5-flowered. *Flowers* with petals blue to whitish. *Capsule* symmetrically 3-locular with the locules dehiscent, smooth, or 2-locular when dorsal locule is aborted, pale cream, shiny, with closely placed horizontal lines; seeds compressed-globose, c. 4 mm long, smooth, farinose, with a long, linear hilum and a semi-circle of large, lighter-coloured cells above embryotega. Fig. 6: 1.

Widespread in the drier parts of tropical and Southern Africa: South West Africa/Namibia, Botswana, Transvaal, N.W. Cape, Orange Free State and Natal; in grasslands and dry bushveld in summer rainfall region. Map 30.

Vouchers: *Burtt-Davy* 2348; *De Winter* 7497; *Esterhuysen* 2228; *Leistner* 1253; *Liebenberg* 108; *Merxmüller & Giess* 1057; *Rodin* 3519; *Schlechter* 3757.

14. Commelina erecta *L.*, Sp. Pl. 1: 41 (1753); Dill., Hort. eltham., tab. 77, fig. 88. Type: to be decided.

C. undulata R. Br., Prodr. 270 (1810); C. B. Cl. in A. DC., Monogr. Phan. 3: 179 (1881) excl. var. *setosa* C. B. Cl.; Rolla Rao in Notes R. bot. Gdn Edinb. 26: 351 (1966). Syntypes: N. Australia, *R. Brown* 5735, 5736 (both BM!, PRE, photo.!).

C. bainesii C.B. Cl. in A. DC., Monogr. Phan. 3: 184 (1881), and in F.T.A. 8: 57 (1901). Type: Zimbabwe, S.W. of Bulawayo, *Baines* (K!).

C. gerrardii C.B. Cl. in A. DC., Monogr. Phan. 3: 183 (1881), and in F.C. 7: 11 (1897). Type: Natal, *Gerrard* (K, holo.).

Perennials (chamaephytes), erect or spreading and rooting at nodes; roots hard, when young covered with a velamen of roothairs. *Leaves* narrowly ovate-attenuate, 60–120 mm long, pseudo-petiolate at junction of lamina with apex of auriculate, ciliate sheath, margins minutely white-pustulate, glabrous or puberulous. *Spathe* pedunculate and solitary or clustered apically, fused, broadly ovate-acute, 15–30 mm long, glabrous to puberulous. *Lower cyme* absent. *Flowers* with petals blue. *Capsule* with 3 one-seeded locules; dorsal locule indehiscent, tuberculate, seed fused to wall; ventral locules dehiscent, smooth; seeds globose smooth, farinose. Fig. 6: 4.

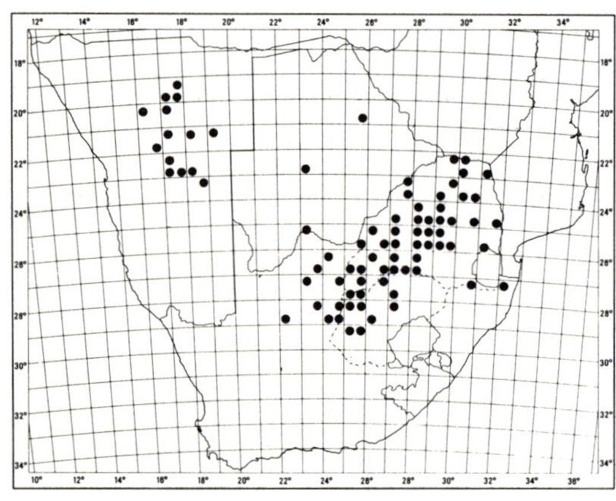

MAP 30.— **Commelina livingstonii**

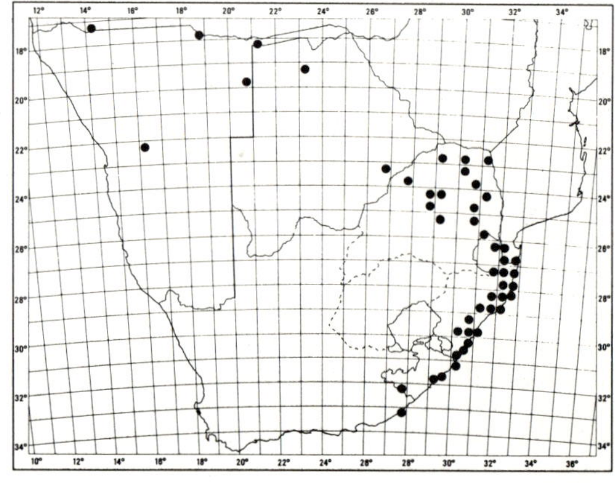

MAP 31.— **Commelina erecta**

A species first described from N. America; common in Asia and Africa, coming as far south as Natal and the E. Cape, where it is common on dunes or sandy flats near the sea; also found in wet habitats, e.g. in spray of Ruacana Falls in South West Africa/Namibia. The spathes are single and shortly pedicelled in typical plants but, especially along the Natal coast, they become clustered and sessile. Map 31.

Vouchers: *De Beer* 577; *De Winter* 3989; *Galpin* 3409; *Kotze* 79; *Pott* 5562; *Seydel* 966; *Strey* 4511; *Vorster* 2578.

In this species, as in *C. forskaolii* (no. 7), the dorsal indehiscent locule of the ripe capsule breaks away with the included seed and part of the adjoining walls of the 2 ventral locules, thus setting free their seeds as well. The ventral locule walls attached to the dorsal locule take on the shape of a shallow cup.

15. **Commelina modesta** *Oberm.* in Bothalia 13: 347 (1981). Type: Transvaal, Barberton, lower hill slopes, *Galpin* 808 (PRE, holo.!).

Small spreading, diffusely branched, glabrescent bushes (chamaephytes) up to c. 0,3 m tall. *Rootstock* woody, gnarled, knobbly (the knobs presenting remains of swollen bases of annual stems); roots woody, long, initially covered by roothairs. *Stems* several, erect, with long internodes up to 60–100 mm long, 1–2 mm in diam. *Leaves* with lamina linear, flat, 80–100 × 4–10 mm, attenuated at base into a pseudo-petiole, white-punctulate; sheath membranous, sub-auriculate. *Flowering spathe* solitary (rarely 2), terminal, sessile or nearly so, fused, shortly triangular, 15 mm long, 10 mm broad, apex short, acute, minutely puberulous and with scattered white setae. *Cyme* solitary. *Flowers* small, petals blue or white ("pink" fide *Galpin* 808). *Sepals* ovate, c. 5 mm, membranous, upper minute. *Petals:* upper ones rounded, c. 15 mm, lower one minute. *Stamens* typical, the anthers occasionally with dark margins, the central semicircular; staminodes with yellow, bulbous antherodes. *Capsule* with globose, shiny, cream-coloured locules; seeds globose, 5 mm in diam., smooth, farinaceous, dorsal seed occasionally aborted.

Widespread in Transvaal, Natal and Swaziland, also in Transkei and eastern Cape; in rocky habitats. Flowering November–March. Map 32.

Vouchers: *Compton* 29564; *Lang* sub TRV 32156; *Moll & Pooley* 4197; *Strey* 10343; *Wild* 7624.

This species was usually placed under *C. livingstonii* (no. 13), but is a more slender bush found in rocky habitats. The leaves narrow gradually below into a pseudo-

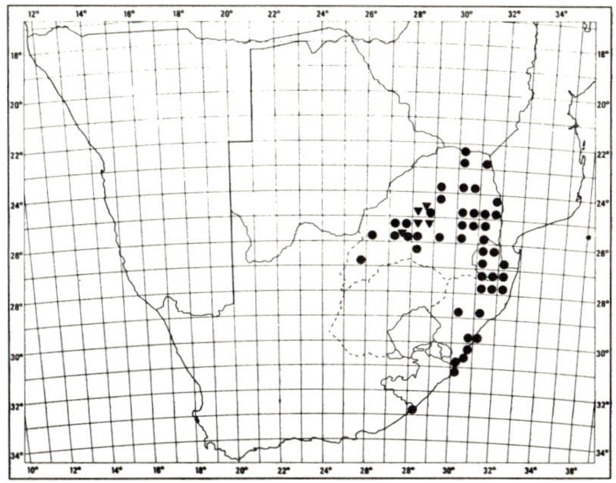

MAP 32.— ● **Commelina modesta**
▼ **Commelina bella**

petiole, whereas in *C. livingstonii* they widen below and then narrow abruptly into the sheath.

16. **Commelina bella** *Oberm.* in Bothalia 13: 436 (1981). Type: Transvaal, 20 km S. of Warmbaths, on Great North Road, along grassy roadside, *Smook* 1494 (PRE, holo.!).

Sturdy erect, compact bushes c. 0,35 m tall, setulose, setae short, white; chamaephytes with a hard gnarled root-crown and woody roots. *Stems* erect, firm, setulose or smooth, internodes c. 50 mm long. *Leaves* folded, linear, c. 50 × 10 mm, apex cirrhose to acuminate-recurved, base merging into a short, open, light greyish green sheath, margins undulate, forming a thick white rim. *Spathes* terminal, opposite the upper leaves, subsessile, fused, triangular, apex acuminate, recurved, c. 25 mm long. *Cymes* 2, upper much exserted on a hairy peduncle c. 25 mm long, with one male flower, petals pale blue to lilac, large; lower cyme c. 3-flowered, flowers bisexual, petals c. 20 mm; stamens 3, two normal with blue locules, central curled up with a large connective, filaments white; staminodes with purple filaments and orange-yellow antherodes consisting of 4 bulbous organs and 2 smaller ones; ovary narrowly ovoid, style exserted below. *Capsule* oblong-globose, hard, c. 10 mm long, 3-seeded; seeds smooth (no normal ripe capsule seen).

Recorded from the Springbok Flats in the Transvaal; usually in turf soil. Map. 32.

Vouchers: *Emmenis* sub PRE 38063; *Mauve* 4279; *Pole Evans* 3852.

899 **2. ANEILEMA**

Aneilema *R. Br.*, Prodr. 270 (1810); C. B. Cl. in A. DC., Monogr. Phan. 3: 195 (1881), and in F.C. 7: 12 (1897), and in F.T.A. 8: 62 (1908); Brückner in Bot. Jb., Beibl. 137: 62 (1926); Brenan in Kew Bull. 7: 180 (1952); Morton in J. Linn. Soc., Bot. 59: 431 (1966); Faden in Agnew, Upland Kenya Wild Flow. 664 (1974), and in Bothalia 15: 89 (1984); R. A. Dyer, Gen. 2: 909 (1976). Type species: *A. biflorum* R. Br.

Perennial or annual herbs of various habits with fibrous or tuberous roots. *Leaves* petiolate or sessile, distichous or spirally arranged. *Inflorescences* thyrses (sometimes reduced to a single cyme), terminal or terminal and axillary, rarely all axillary. *Bracteoles* persistent, usually cup-shaped and perfoliate. *Flowers* bisexual, bisexual and staminate, or bisexual, staminate and pistillate, all types produced in same inflorescence. *Sepals* 3, free, predominantly green. *Petals* 3, free, upper 2 (paired petals) clawed, lower one (medial petal) usually reduced and discolorous (rarely subequal and concolorous, but then different in form). *Staminodes* 3, posticous (on the upper side), filaments glabrous, antherodes bilobed. *Stamens* 3, anticous (on the lower side), medial (antepetalous) one different in form and size from lateral (antesepalous) ones, filaments all glabrous or laterals sparsely to densely bearded. *Ovary* bi- or trilocular, dorsal locule suppressed or 1-ovulate, ventral locules each with 1–6 uniseriate ovules; style simple, stigma capitate or not enlarged. *Capsule* usually dehiscent and bivalved (rarely ± indehiscent), dorsal locule suppressed or 1-seeded, ventral locules each 1–6-seeded. *Seeds* with a linear hilum and lateral embryotega.

Species about 60, mainly tropical Africa; in Southern Africa 11 species, mainly in eastern and tropical regions.

The name Aneilema (= without covering) refers to the absence of a spathe.

1 Plants prostrate, mat-forming; inflorescences all axillary, perforating the sheaths, consisting of 1(–2) cyme(s) ... . 6. *A. zebrinum*
1 Plants (or at least the flowering shoots) erect to ascending or decumbent; some or all inflorescences terminal, thyrsiform, composed of several to many cymes:
  2 Bracteoles not cup-shaped, apex drawn out into an elongate, swollen, glandular tip; lateral stamen filaments densely bearded with blue-purple hairs; annuals .... 11. *A. nicholsonii*
  2 Bracteoles cup-shaped, apex not drawn out into an elongate, glandular tip; lateral stamen filaments glabrous or bearded with white hairs; annuals or perennials:
    3 Flowers yellow to orange; inflorescences lax:
      4 Leaves spirally arranged; flowers orange-yellow; inflorescences with c. 10–20 cincinni; inflorescences, sepals and stamen filaments glabrous .... 4. *A. johnstonii*
      4 Leaves distichous; flowers yellow; inflorescences with 2–7 cincinni; inflorescences and sepals puberulous, lateral stamen filaments bearded .... 3. *A. aequinoctiale*
    3 Flowers pink to blue-purple (or white); inflorescences lax to dense:
      5 Perennials with thick roots; flowers 13,5–40 mm wide; lateral stamen filaments undulate, glabrous; capsules 8–14 mm long:
        6 Leaf margins scabrid; branches of the inflorescence mostly subopposite or subverticillate; stamen filaments fused basally; fruiting pedicels erect; capsules emarginate at the apex .... 1. *A. hockii*
        6 Leaf margins smooth; branches of the inflorescence mostly alternate; stamen filaments free; fruiting pedicels decurved; capsules rounded to truncate at the apex .... 2. *A. longirrhizum*
      5 Perennials or annuals with thin roots; flowers 6,5–17,5 mm wide; lateral stamen filaments sigmoid, bearded or glabrous; capsules c. 3–7,5 mm long:
        7 All 3 petals similar in colour and size; stamen filaments glabrous; capsules trilocular .... 5. *A. indehiscens*
        7 Lower petal distinctly smaller than the upper 2 and different in colour; lateral stamen filaments bearded; capsules bilocular:
          8 Perennials; capsules oblong-elliptic to obovate-oblong or oblong, (3,5–)5–7,5 mm long, locules 2-seeded .... 7. *A. dregeanum*
          8 Annuals; capsules broadly elliptic to obovate or obovate-orbicular, 2,8–4,5(–5) mm long, locules 1-seeded:
            9 Pedicels 1,5–3 mm long, ± erect in fruit; capsules with valves relatively planar; cells of the outer capsule wall (with 20× lens) transversely elongate; seeds buff or light brownish orange .... 8. *A. arenicola*

9 Pedicels 2–8 mm long, strongly recurved in fruit; capsules with valves strongly convexo-concave; cells of the outer capsule wall longitudinally elongate or ± isodiametric; seeds dark brown or pale pinkish grey:
  10 Inflorescences composed of (6–)10–20(–29) cymes; pedicels puberulous for up to half their length; cells of the capsule wall longitudinally elongate; seeds dark brown . . . . **9. *A. brunneospermum***
  10 Inflorescences composed of 3–13 cymes; pedicels puberulous for more than half their length; cells of the capsule wall ± isodiametric; seeds pale pinkish grey . . . . **10. *A. schlechteri***

1. **Aneilema hockii** *De Wild.* in Feddes Reprium 12: 290 (1913); Brenan in Kew Bull. 7: 190 (1952); Schreiber et al. in F.S.W.A. 157: 2 (1967); Faden in Agnew, Upland Kenya Wild Flow. 664 (1974). Type: Zaire: Upper Katanga, Elisabethville, 1911, *Hock* s.n. (BR!).

Aneilema aequinoctiale (P. Beauv.) Loudon var. *kirkii* C. B. Cl. in A. DC., Monogr. Phan. 3: 222 (1881). Syntypes: Mozambique: Chupanga (Shupanga), in damp spots, 10 January 1863, *Kirk* s.n. (K!, lectotype of Brenan, l.c., 193); same locality, January 1859, *Kirk* s.n. (K!); Near Sena (Senna), 3 January 1860, *Kirk* s.n. (K!).

Aneilema aequinoctiale auct. p. p., e.g. Schinz in Bull. Herb. Boissier 4, App. 3: 36 (1896), p. p.; C. B. Cl. in F.C. 7: 12 (1897), p. p., and in F.T.A. 8: 65 (1901), p. p.; Van Druten in Flow. Pl. Afr. 33: 1302 (1959), non (P. Beauv.) Loudon (1830).

Aneilema wildii Merxm. in Suessenguth & Merxmüller, Trans. Rhod. Sci. Assoc. 43: 152 (1951). Type: Zimbabwe: Marandellas, 14 January 1941, *Dehn* 251 (M!).

Tufted perennial with erect to ascending or straggling shoots 0,3–1,2 m tall. *Roots* thick, fleshy. *Leaves* spirally arranged, leaf-blades sessile or petiolate, usually linear-lanceolate to lanceolate or oblong-lanceolate, (30–)50–150(–245) × (3–)10–35(–48) mm, apex usually acuminate. *Inflorescences* mostly terminal, (25–)40–105 × 35–100(–120) mm, with 4–11(–16) mostly opposite or whorled cincinni. *Flowers* perfect and staminate, (15–)20–40 mm wide; paired petals bluish purple, mostly c. 15 × 15 mm; lateral stamen filaments c. 15–30 mm long, glabrous; style c. 15–25 mm long. *Capsules* oblong to oblong-elliptic, bilocular, 7–13 × 4–5,5 mm, locules 3–5-seeded. *Seeds* transversely rectancular to trapezoidal or ovate, 1,5–3,25 × 2,15–2,85 mm, testa shallowly scrobiculate to slightly rugose. Fig. 7: 1.

South West Africa/Namibia, Botswana, Transvaal and Swaziland and north to Zaire, Uganda and southern Ethiopia; c. 640–1 280 m altitude (in our area); woodland and bushland, often on rocky slopes; growing in partial shade. Map 33.

Vouchers: *De Winter & Leistner* 5599; *Faden & Faden* 74/215; *Pole Evans* 2593; *Schlechter* 4565; *Thorncroft* 265.

The large, bluish purple flowers (variously described as blue, mauve, lavender, etc.) readily distinguish this species. The capsules are generally longer, less pubescent (commonly nearly glabrous) and have more seeds per locule than those of *A. aequinoctiale* (no. 3). Typically the capsules are emarginate and lack a terminal apicule (the persistent style base), unlike the capsules of *A. aequinoctiale.*

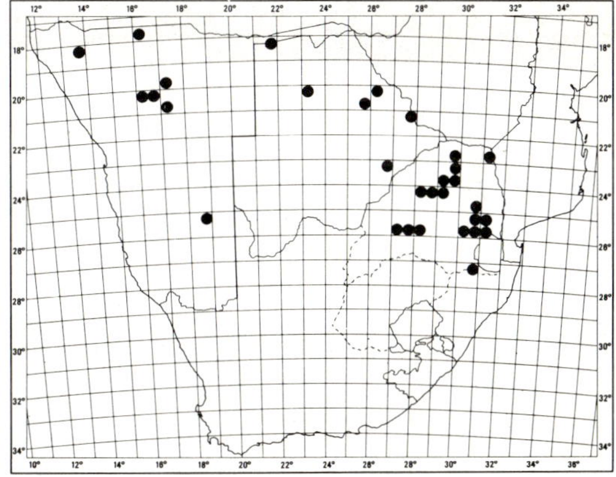

MAP 33.— **Aneilema hockii**

2. **Aneilema longirrhizum** *Faden* in Bothalia 12: 565 (1978); Flow. Pl. Afr. 45: 1785 (1979). Type: Transvaal, Burgersfort–Pietersburg road, 3,4 km towards Pietersburg from crossing of the Olifants River, 24 February 1974, *Faden & Faden* 74/217 (K, holo.!; B!; BOL!; BR!; C!; EA!; F!; FI!; G!; HBG!; M!; MO!; NH!; NU!; P!; PRE!; S!; SRGH!; UPS!; WAG!. iso.).

Rhizomatous perennial. *Roots* thick, fleshy. *Leaves* distichous or spirally arranged, leaf-blades sessile, linear-lanceolate to narrowly lanceolate-elliptic, 30–130 × 6–20 mm. *Inflorescences* terminal, ovoid thyrses 40–130(–180) mm long, with (6–)8–15(–24) mostly alternate cincinni. *Flowers* perfect and staminate, fragrant, 13,5–18 mm wide; pedicels to 9 mm long, recurved c. 270° (i.e. pointing downwards) in fruit; sepals puberulous, 3,7–4,5 mm long; paired petals pale lavender, 7,5–11,5 × 6–9 mm, medial petal 5,5–7 mm long; lateral stamen filaments 8,5–9,5 mm long; style 8,5–12 mm long. *Capsules* oblong-elliptic, bilocular, 8–11 ×

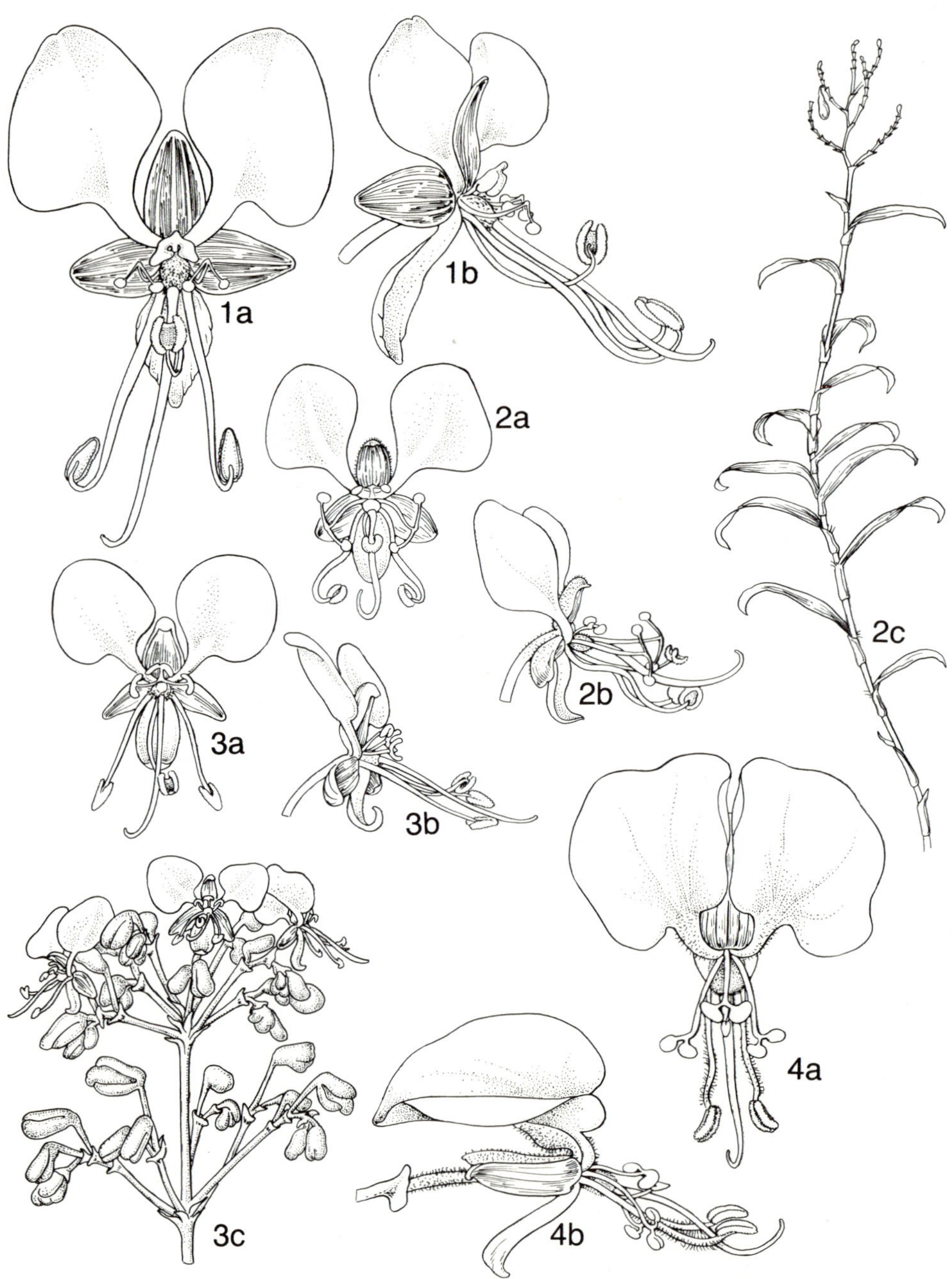

Fig. 7.—1, **Aneilema hockii**: 1a, flower, front view, × 2; 1b, flower, side view, × 2 (*Faden & Faden* 74/190). 2, **A. longirrhizum**: 2a, flower, front view, × 2; 2b, flower, side view, × 2; 2c, habit, × 0,25 (*Faden & Faden* 74/217). 3, **A. johnstonii**: 3a, flower, front view, × 2; 3b, flower, side view, × 2; 3c, inflorescence, × 1 (*Pawek* 12327). 4, **A. aequinoctiale**: 4a, flower, front view, × 2; 4b, flower, side view, × 2 (*Faden & Faden* 74/199).

4–5 mm, apex rounded to truncate, locules 2(–1)-seeded. *Seeds* 3,4–4,9 mm long, densely white-farinose. Fig. 7:2.

Local endemic in northern Transvaal, 760–850 m altitude. Dry habitats; recorded from *Blepharis-Commiphora-Schmidtia* bushed grassland on red loamy soil, and outcrops and residual soils of the Old Granite. Map 34.

Vouchers: *Bremekamp & Schweickerdt* 404; *Faden & Faden* 74/217; *Matthie* 567 (PRU); *Mogg* 1156.

This species can be confused only with *A. hockii* (no. 1) which differs, in addition to the characters mentioned in the key, by its larger flowers, striped sepals with hook-hairs of two size classes, differently shaped staminodes, and smaller, more numerous seeds (per locule) which lack farinose granules.

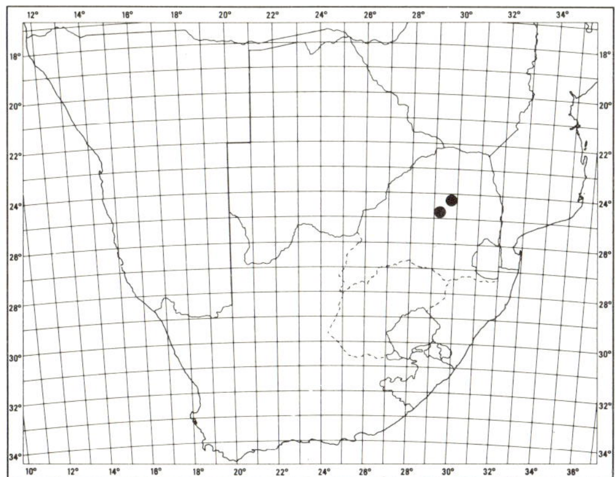

MAP 34.— **Aneilema longirrhizum**

3. **Aneilema aequinoctiale** *(P. Beauv.) Loudon*, (Loudon's) Hort. Brit., 15 (1830); Kunth, Enum. Pl. 4: 72 (1843); C. B. Cl. in A. DC., Monogr. Phan. 3: 221 (1881), p. p. and in F.C. 7: 12 (1897), p. p. and in F.T.A. 8: 65 (1901), p. p.; Brenan in Kew Bull. 7: 194 (1952); Morton in J. Linn. Soc., Bot. 59: 443 (1966); Brenan in F.W.T.A., edn 2, 3: 30 (1968); Ross, Fl. Natal, 117 (1972); Faden in Agnew, Upland Kenya Wild Flow., 664 (1974); Gibson, Wild Flow. Natal, Pl. 4, fig. 5 (1975); Compton, Fl. Swaziland, 83 (1976). Type: Nigeria: Oware and Benin, without specific locality. *Palisot de Beauvois* s.n. (G!).

Commelina aequinoctialis P. Beauv., Fl. Oware 1: 65, t. 38 (1806). *Lamprodithyros aequinoctialis* (P. Beauv.) Hassk. in Schweinfurth, Beitrag Fl. Aethiop., 211 (1867).

Aneilema adhaerens Kunth, Enum. Pl. 4: 72 (1843). *Lamprodithyros adhaerens* (Kunth) Hassk. in Schweinfurth, Beitrag Fl. Aethiop., 211 (1867). *Aneilema aequinoctiale* (P. Beauv.) Loudon var. *adhaerens* (Kunth) C. B. Cl. in A. DC., Monogr. Phan. 3: 222 (1881). Type: Cape, St. John's River, 1839, *Drège* 4466 (G, lecto.!; B; FHO!; MO!; P!; S!).

Decumbent perennial with ascending or straggling shoots. *Roots* thin, fibrous. *Leaves* distichous, sheaths "sticky", leaf-blades (except the upper) petiolate, usually lanceolate-elliptic to ovate, 35–130(–165) × (10–)15–40(–50) mm, apex acute to acuminate. *Inflorescences* mostly terminal, 30–115 × 20–70 mm, with 2–7 alternate, opposite or whorled cincinni. *Flowers* perfect and staminate, c. 18–30 mm wide; paired petals yellow, c. 15 × 15 mm; lateral stamen filaments c. 11–18 mm long, bearded; style 11–21 mm long, purple. *Capsules* oblong-elliptic to obovate-oblong or obovate, trilocular or bilocular, (5–)7–10 × 3,5–6 mm, dorsal locule 1(–0)-seeded, ventral locules 2–3-seeded. *Seeds* mostly ovate to trapezoidal or subquadrate, 1,9–2,6 × 1,95–2,15 mm, testa brown, faintly to shallowly foveolate-reticulate. Fig. 7: 4.

Transvaal, Swaziland, Natal and Transkei; also throughout tropical Africa to Guinea and Ethiopia. Grows at c. 0–1 000 m (–2 000 m?) in our area, in moist places, especially forests, forest edges and along streams, also roadsides, bush, thickets and occasionally grassland, usually in partial shade. Map 35.

Vouchers: *Bos* 1235; *Compton* 27284; *Faden & Faden* 74/199; *Galpin* 3193; *Strey* 7074.

The stems and sheaths often feel sticky because of the presence of hooked hairs. The plant is sometimes described as scrambling or scandent because the long shoots often become entangled in other plants. *Aneilema aequinoctiale* is easily separated from *A. johnstonii* (below) by its larger flowers and fewer-branched, pubescent inflorescences. When flowers are not available, it can be confused with *A. hockii* (no. 1), which differs by its more tufted habit, proportionally narrower leaves, usually more branched inflorescences in which the branches (cincinni) are mostly whorled, and usually non-apiculate capsule valves. Under high magnification the cells of the capsule

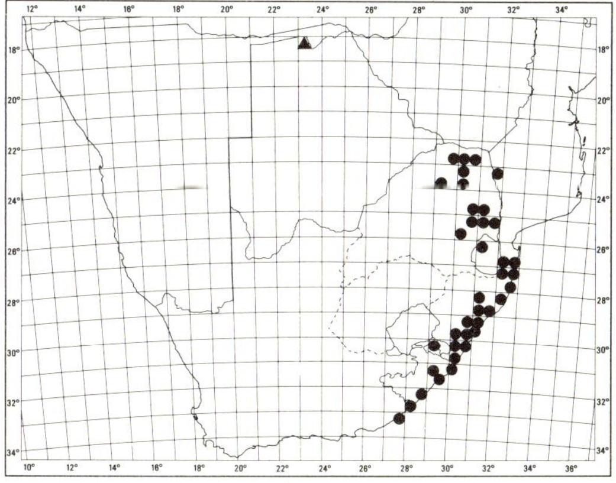

MAP 35.— ● **Aneilema aequinoctiale**
      ▲ **Aneilema johnstonii**

wall of *A. aequinoctiale* are isodiametric whereas those of *A. hockii* are transversely elongate.

4. **Aneilema johnstonii** *K. Schum.* in Engl. Pflanzenw. Ost-Afr. C, 135 (1895); C. B. Cl. in F.T.A. 8: 67 (1901); Faden in Agnew, Upland Kenya Wild Flow. 664 (1974). Syntypes: Tanzania: Kilimanjaro, October, 1884, *Johnston* s.n. (B; BM!; K!); West of Pare, 1887–88, *von Höhnel* 159 (B!); Kilimanjaro, below Marangu, 24 April 1984, *Volkens* 2146 (B, lecto.!; BM, isolecto.!).

Tufted perennial with annual flowering shoots to c. 600 mm tall. *Roots* with distal, fusiform tubers. *Leaves* spirally arranged, leaf-blades sessile or petiolate, linear-lanceolate to lanceolate (or lanceolate-elliptic), 50–150 × (6–)10–25(–33) mm, apex usually acuminate. *Inflorescences* terminal, glabrous, c. 40–100 × 35–70 mm, with mostly 10–20 cincinni arranged in 2–6 whorls. *Flowers* perfect and staminate, c. 12–18 mm wide; paired petals orange-yellow, 7–9,5 mm long; lateral stamen filaments 9–12 mm long, glabrous; style 10–13 mm long. *Capsules* elliptic to obovate, trilocular or bilocular, (3,5–)4–6 × 3–4,5 mm, dorsal locule 1(–0)-seeded, ventral locules 1–2(–3)-seeded. *Seeds* of ventral locules elliptic to ovate, 1,8–3,2 × 1,6–2,2 mm, testa smooth to faintly reticulate. Fig. 7:3.

Botswana; also Mozambique and Zimbabwe to southern Ethiopia; c. 950 m altitude (in our area). Woodland, grassland or bushland, sometimes thickets, often in rocky places. Map 35.

Vouchers: *Curson* 747; *Henry* 34 (SRGH).

The completely glabrous inflorescences of *A. johnstonii* are unique among our species and readily separate this plant from *A. aequinoctiale* (above), the only other yellow-flowered *Aneilema* in the area. Furthermore, the two species occur in different habitats and do not overlap geographically in Southern Africa. The subequal, horseshoe-shaped anthers of the staminodes in *A. johnstonii* are highly distinctive.

5. **Aneilema indehiscens** *Faden* subsp. **lilacinum** *Faden* in Bothalia 15: 97 (1984). Type: Natal, Ingwavuma-Ndumu road, 15,5 km towards Ingwavuma from junction with Ndumu-Maputa road, c. 27°06′S, 32°12′E, 16 February 1974, *Faden & Faden* 74/202 (US, holo.!; BR!; EA!; K!; LISC!; MO!; NH!; NU!; PRE!; US!; WAG!, iso.).

Aneilema dregeanum sensu Compton, Fl. Swaziland, 83 (1976), p. p., non Kunth (1843).

Perennial with long-trailing vegetative shoots. *Roots* thin, fibrous. *Leaves* spirally arranged, leaf-blades shortly petiolate, narrowly lanceolate to lanceolate-elliptic, lanceolate-ovate or ovate-elliptic, (25–)30–100(–130) × (7–)10–25(–35) mm. *Inflorescences* terminal or terminal and axillary, ovoid thyrses (20–)25–50(–80) × (15–)20–50(–70) mm with (1–)3–9 mostly ascending cincinni. *Flowers* perfect and staminate, (9–)13–17,5 mm wide; pedicels recurved usually c. 180° in fruit; petals pale lilac, medial petal large, cup-shaped, usually obovate; lateral stamen filaments usually ± parallel, 7,7–8,5 mm long, glabrous. *Capsules* obovate-elliptic to obovate-oblong, oblong or oblanceolate, trilocular, dehiscent, (4–)4,5–6(–6,8) × (1,9–)2,3–3(–3,4) mm, dorsal locule 1-seeded, ventral locules 2-seeded. *Seeds* elliptic or ovate to trapezoidal, 1,5–2,9 × 1,35–2,2(–2,5) mm, testa shallowly scrobiculate. Fig. 9:·2.

Northern Transvaal, Swaziland and northern Natal; also southern Zimbabwe and southern Mozambique; c. 10–550 m altitude. Open forest, woodland, thickets, lowveld bush and edges of marshes in sandy or clayey soils, usually in partial shade. Map 36.

Vouchers: *Codd* 6891; *Faden & Faden* 74/202; *Faden & Faden* 74/208; *Moll* 4152; *Strey* 10326.

This species is readily distinguishable from *A. dregeanum* (no. 7) and *A. brunneospermum* (no. 9), with which it has been confused, by its laxer inflorescences with fewer cincinni, large, cup-shaped medial petal, glabrous stamen filaments, and trilocular ovaries and capsules. Subsp. *indehiscens* is restricted to eastern Kenya and northeastern Tanzania.

6. **Aneilema zebrinum** *Chiov.* in Webbia 8: 38, fig. 12, p. 39 (1951). Syntypes: Ethiopia, Gemu-Gofa Prov., rive del Caschei, 5

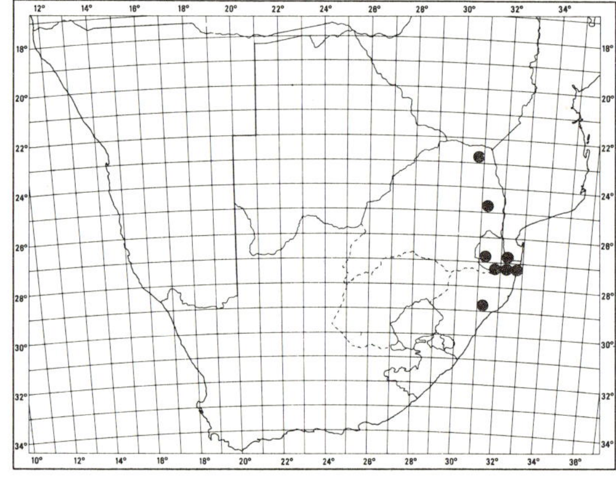

MAP 36.— **Aneilema indehiscens** subsp. **lilacinum**

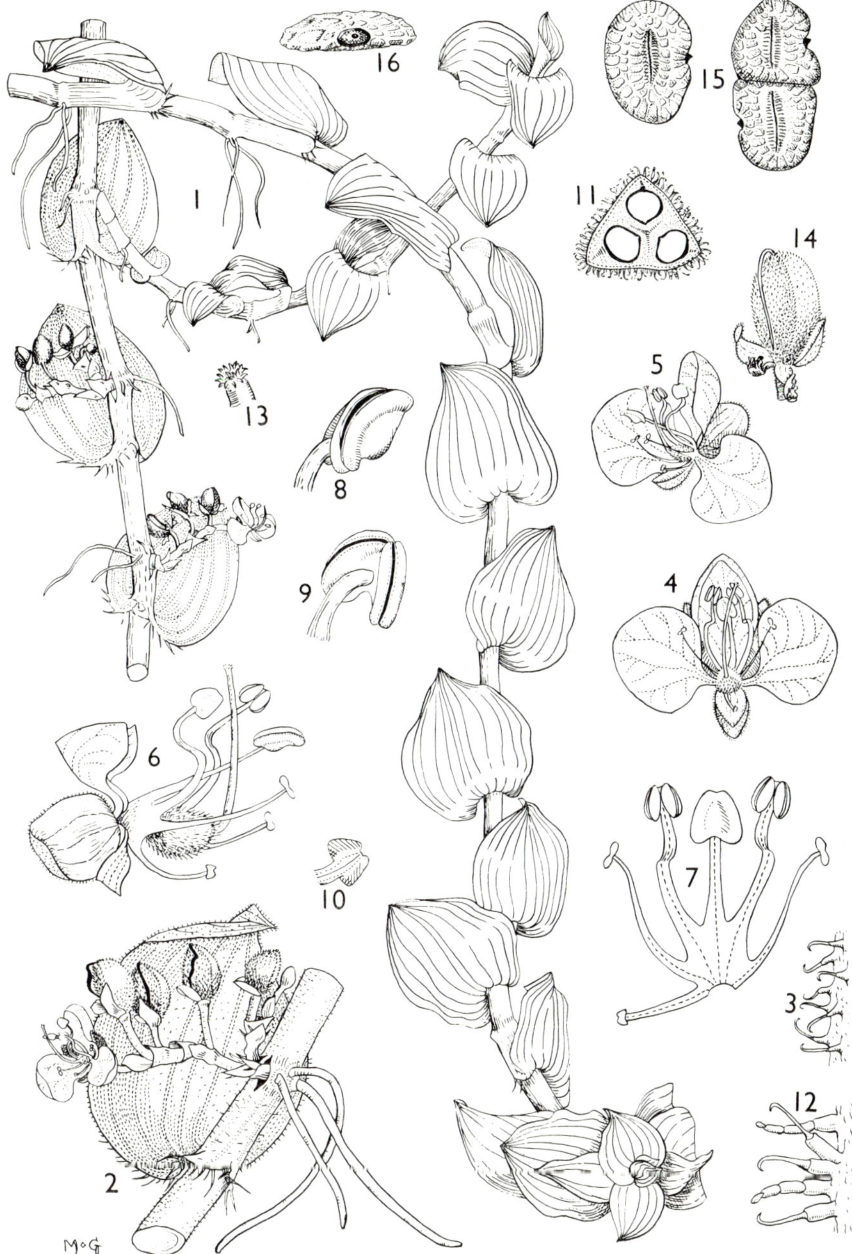

FIG. 8.—**Aneilema zebrinum**: 1, part of flowering plant, × 1; 2, single node of plant, showing adventitious roots and inflorescence perforating the leaf-sheath, × 2; 3, hairs on stem (similar ones on sepals), × 60; 4, flower from front, × 4; 5, flower from side, × 4; 6, flower from side with sepals and petals partly removed, × 9; 7, androecium, split on one side of central staminode and spread out, × 9; 8, fertile anther, from side, × 16; 9, fertile anther, from back, × 16; 10, apex of staminode, × 20; 11, ovary, cross-section, × 18; 12, hairs on surface of ovary, × 60; 13, stigma and apical part of style, × 24; 14, fruit, before dehiscence, × 4; 15, seeds, from a one-seeded loculus to left, from a two-seeded loculus to right, × 9; 16, seed, from embryotegal edge, × 9. All from *Bally* 12145 (cultivated). Reproduced, from 'Kew Bulletin' with permission of the Director of the Royal Botanic Gardens, Kew.

July 1939, *Corradi* 2154 (FI!, K, photo.!); same locality, 4 July 1939, *Corradi* 2163 (FI!, K, photo.!); Caschei, 6 July 1939, *Corradi* 2155 (FI!; lectotype of Brenan in Kew Bull. 19: 67 (1964); K, photo.!).

Ballya zebrina (Chiov.) Brenan in Kew Bull. 19: 64, fig. 1, p. 66 (1964); Cufodontis in Bull. Jard. bot. nat. Belg. 41, Suppl., 1519 (1971); Ross, Fl. Natal, 117 (1972); Faden in Agnew, Upland Kenya Wild Flow. 653 (1974).

Repent, often mat-forming perennial. *Roots* thin, fibrous. *Leaves* distichous, sessile, succulent, ovate to ovate-elliptic or occasionally elliptic, 10–35(–40) × 5–20 mm, puberulous, often mottled with maroon above, veins contrastingly pale above. *Inflorescences* axillary, perforating the sheaths and largely hidden beneath leaves, consisting of 1(–2) short, few-flowered cincinnus. *Flowers* perfect, 7–10 mm wide; petals all pale lilac, paired petals 4–5,7 × 3,1–4,5 mm, medial petal cup-shaped, 3,5–4,7 × 2,7–4 mm; filaments fused basally, lateral stamen filaments 3–3,5(–4,5) mm long; style 2,7–3,2(–3,6) mm long. *Capsules* obovate-elliptic, indehiscent or partially dehiscent, trilocular, 3–4 × 2,1–2,5(–3) mm, dorsal locule 1-seeded, ventral locules 2-seeded. *Seeds* 1,25–1,8 mm long, testa shallowly reticulate. Figs 8 & 9: 3.

Northern Natal to southwestern Ethiopia; c. 25–30 m altitude (in the Flora area); thickets and open forest. Map 37.

Vouchers: *Faden et al.* 74/205; *Pooley* 1255 (K, MO); *Strey* 4752.

This species is unmistakable because of its prostrate habit and strictly axillary inflorescences. Its unusual features led to its being described as a distinct genus by Brenan. However, it is connected to the rest of *Aneilema* by a series of local, endemic taxa from Tropical East Africa.

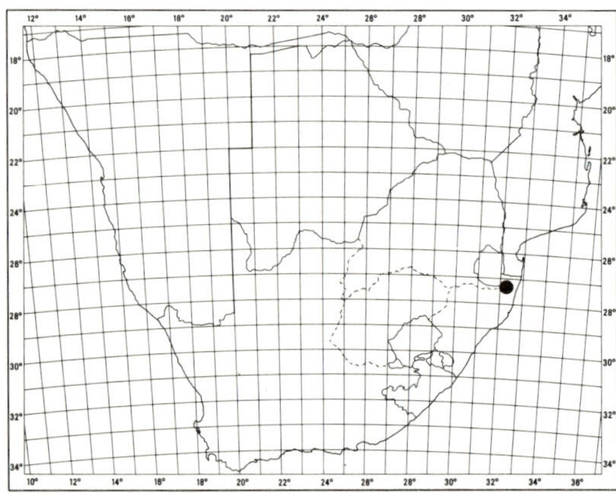

MAP 37.— **Aneilema zebrinum**

The low stature and inconspicuous flowers of *A. zebrinum* render it easily overlooked. Its relatively recent discovery in South Africa was a remarkable range extension from northern Tanzania, the then nearest locality. The plant has not been collected in Mozambique, but it is almost certainly widespread there.

7. **Aneilema dregeanum** *Kunth*, Enum. Pl. 4: 73 (1843); C. B. Cl. in A. DC., Monogr. Phan. 3: 229 (1881), and in F.C. 7: 13 (1897); Brenan in Kew Bull. 15: 215 (1961); Faden in Bothalia 15: 90 (1984). Type: Pondoland, between Umtata River and St. Johns River, 1839, *Drège* 4471 (B, holo.!; B!; BM; FHO!; G!; K!; MO!; P!; S, iso.).

subsp. **dregeanum.**

Decumbent perennial. *Roots* thin, fibrous. *Leaves* spirally arranged, leaf-blades with long to fairly short petioles, narrowly lanceolate or lanceolate-elliptic to ovate or ovate-elliptic, (20–)30–110(–150) × (6–)10–45 mm. *Inflorescences* terminal or terminal and axillary, moderately dense ovoid thyrses 20–55 × 10–35(–40) mm, with (8–)11–27, mostly alternate, ascending to patent cincinni. *Flowers* perfect and staminate, 9,5–14,5 mm wide; pedicels 3,5–8 mm long; sepals 2,3–4 mm long, sparsely puberulous; paired petals blue to blue-violet, (6–)7–9,5 × 5,5–6,5 mm, medial petal 3–4 mm long; lateral stamen filaments 6–11 mm long, finely bearded apically with minute white hairs; style 7–11 mm long, violet apically. *Capsules* shortly stipitate, bilocular, usually oblong-elliptic, (3,5–)5–7,5 × (2,8–)3,5–4,2 mm, locules 2-seeded. *Seeds* ovate to trapezoidal, 1,95–3 mm long, testa foveolate-scrobiculate. Fig. 10: 3.

Natal, eastern Cape Province; moist or mesic habitats, most commonly in forests, also in bush and, rarely, (derived?) grassland; c. 0–940 m. Map 38.

Vouchers: *Acocks* 13312; *Faden & Faden* 74/214; *Huntley* 63; *Strey* 9484; *Wells* 3493.

The Southern African plants all belong to subsp. *dregeanum*. Subsp. *mossambicense* Faden (Bothalia 15: 90, 1984), which differs by its generally narrower, more shortly petiolate leaves, laxer inflorescences, longer-bearded stamen filaments and rugose to scrobiculate seeds, is confined to northern Mozambique.

8. **Aneilema arenicola** *Faden* in Bothalia 15: 94, f. 2 (1984). Type: Natal, Ubombo–Sordwana Bay road, 2,6 km towards Sordwana Bay from Shongwe, 19 February 1974, *Faden & Faden* 74/211 (US, holo.!; B!; BR!; EA!; K!; MO!; NH!; NU!; P!; PRE!; UPS!, iso.).

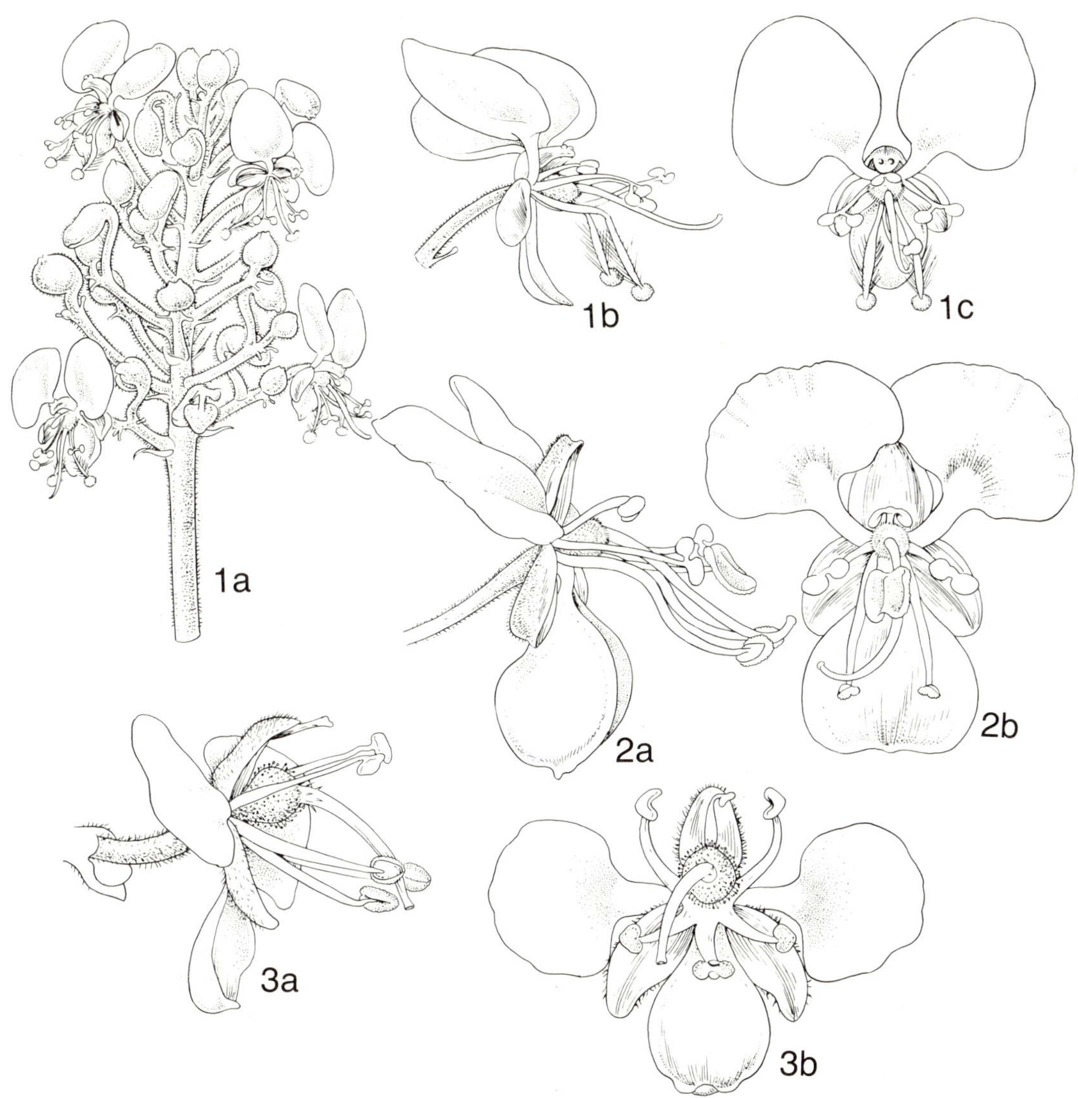

Fig. 9.—1, **Aneilema nicholsonii**: 1a, inflorescence, × 2; 1b, flower, side view, × 3; 1c, flower, front view, × 3 (*Pawek* 12545). 2, **A. indehiscens** subsp. **lilacinum**: 2a, flower, side view, × 3; 2b, flower, front view, × 3 (*Faden & Faden* 74/208). 3, **A. zebrinum**: 3a, flower, side view, × 6, 3b, flower, front view, × 6 (*Faden & Faden & Faulkner* 74/330).

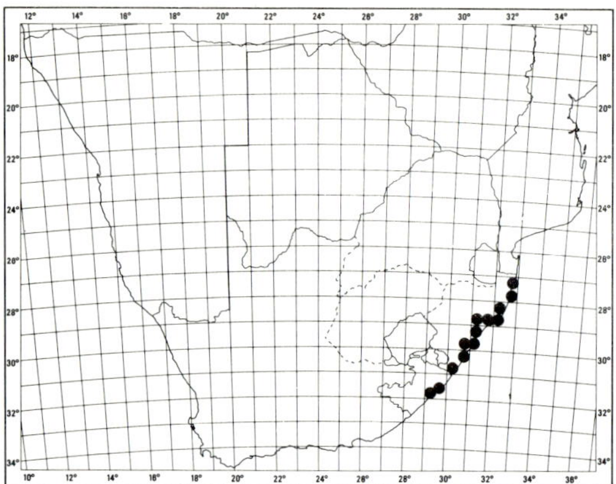

Map 38.— **Aneilema dregeanum** subsp. **dregeanum**

Densely branched annual. *Roots* thin, fibrous. *Leaves* distichous, leaf-blades usually sessile to shortly petiolate, lanceolate-elliptic to ovate, 15–45(–65) × 7–20(–25) mm. *Inflorescences* terminal, ovoid thyrses 20–50 mm long and wide with (2–)5–13 ascending cincinni. *Flowers* perfect, 6,5–8,5(–9,5) mm wide; pedicels 1,5–3 mm long; sepals puberulous, (1,5–)2–3,5(–3,8) mm long, subapical glands bilobed; paired petals pale lilac, 3,5–5 × 2,7–3,5 mm, medial petal 2–2,5(–2,8) mm long; lateral stamen filaments 2,5–3 mm long, densely bearded; style 1,5–2,3 mm long. *Capsules* broadly elliptic to obovate-orbicular, bilocular, 3–4 × 2,6–3,75 mm, locules 1-seeded. *Seeds* elliptic, 2,2–3,1 × 1,5–1,9 mm, testa buff to light brownish orange, foveolate-reticulate.

Northern Natal; also southern Mozambique, 10–60 m altitude. Sandy soil along roadsides and tracks and in partially open woodland. Map 39.

Vouchers: *Faden et al.* 74/204; *Faden & Faden* 74/211; *Moll* 4621; *Pooley* 1679; *Stewart* 1698 (MO).

This species can be confused only with *A. brunneospermum* (below), the only other annual species in its range. *Aneilema arenicola* differs in its ecology, more branched habit, more widely spaced bracteoles, smaller, solely perfect flowers, less firm capsule valves with the outer wall cells transversely elongate, and larger, paler, more dorsiventrally flattened seeds. Fig. 10: 2.

9. **Aneilema brunneospermum** *Faden* in Bothalia 15: 91, fig. 1 (1984). Type: Natal, c. 5 km on road to Ndumu from the junction of the Ingwavuma-Ndumu and Ingwavuma-Ubombo roads, 18 February 1974, *Faden, Faden & Pooley* 74/209 (US, holo.!; B!; BR!; K!; MO!; P!; PRE!; WAG!, iso.).

Aneilema dregeanum Kunth var. *galpinii* C. B. Cl. in F.C. 7: 13 (1897); Brenan in Kew Bull. 15: 216 (1961), pro syn. Type: Transvaal, Barberton (details of specific localities differ on all three specimens), 16 December 1890, *Galpin* 1187 (K, holo.!; NH!; PRE!).

Aneilema schlechteri sensu Brenan in Kew Bull. 15: 216 (1961), p. p.

Aneilema dregeanum sensu Compton, Fl. Swaziland, 83 (1976), p. p., non Kunth (1843).

Tufted annual. *Roots* thin, fibrous. *Leaves* with petioles usually long to short, leaf-blades lanceolate or lanceolate-elliptic to ovate, 25–100(–140) × (6–)10–35(–60) mm, apex acute to acuminate. *Inflorescences* all or mostly terminal, (20–)25–60(–75) × (15–)20–45(–60) mm, with (6–)10–20(–29) ascending cincinni. *Flowers* perfect and staminate, (9–)11–15 mm wide; pedicels (2,2–)2,5–7(–8) mm long, puberulous above the middle; sepals (2–)2,5–3,6(–4,3) mm long; paired petals lavender or pale lilac, 5,3–7,5(–9) mm long; lateral stamen filaments 4,8–9,5 mm long, bearded, anther sacs blue-black; style

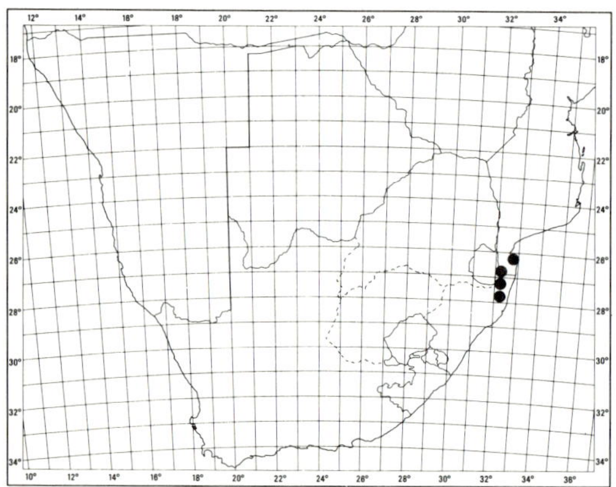

Map 39.— **Aneilema arenicola**

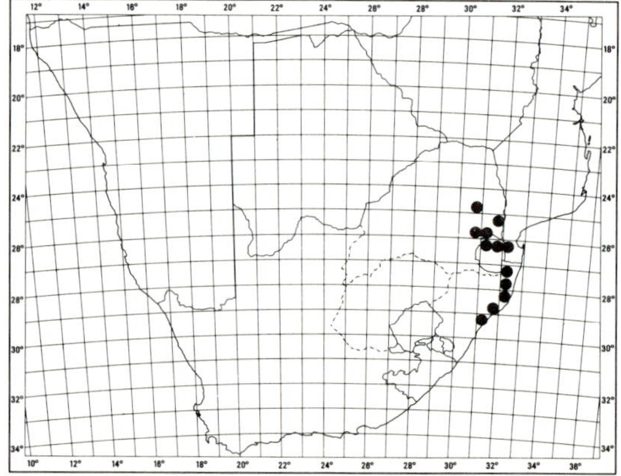

Map 40.— **Aneilema brunneospermum**

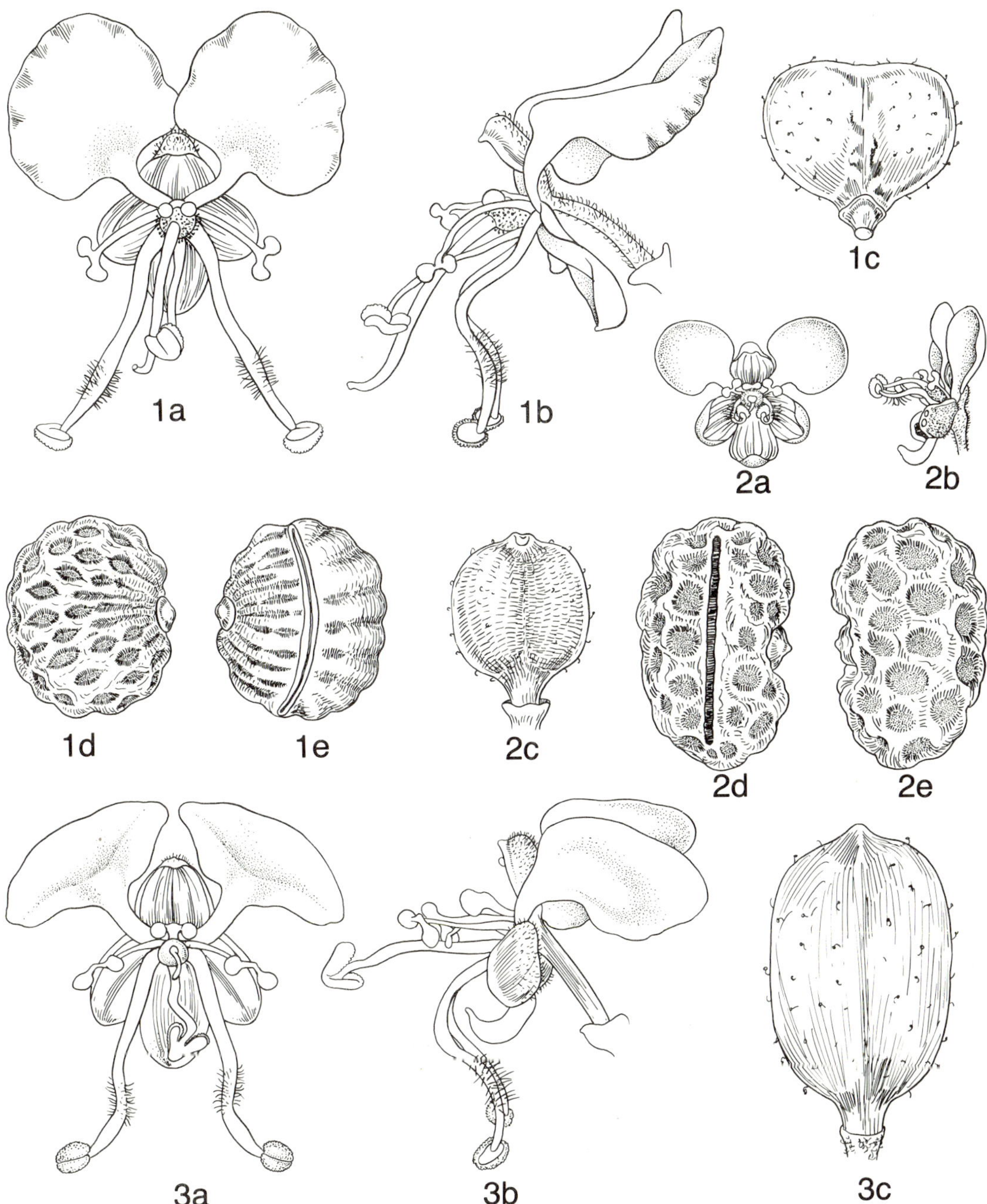

FIG. 10.—1, **Aneilema brunneospermum**: 1a, flower, front view, × 3; 1b, flower, side view, × 3; 1c, capsule, × 6; 1d, seed, dorsal view, × 12,5; 1e, seed, ventral view, × 12,5 (*Faden & Faden & Pooley* 74/209). 2, **A. arenicola**: 2a, flower, front view, × 3; 2b, flower, side view, × 3; 2c, capsule, × 6; 2d, seed, ventral view, × 12,5; 2e, seed, dorsal view, × 12,5 (*Faden & Faden & Pooley* 74/204). 3, **A. dregeanum**: 3a, flower, front view, × 3; 3b, flower, side view, × 3; 3c, capsule, × 6 (3a & 3b from *Faden & Faden* 74/214; 3c from *Bolus* 10348).

(3–)4,5–9 mm long, yellow with violet tip. *Capsules* mostly broadly elliptic to obovate, bilocular, (2,8–)3–4,5(–5) × (3,3–)3,8–4,5(–4,8) mm, locules 1-seeded. *Seeds* elliptic, 2–2,5(–2,8) × 1,7–2,05 mm, testa dark brown, usually foveolate-reticulate. Fig. 10: 1.

Transvaal, Swaziland and Natal; also southern Mozambique. Scrub or forest (rarely grassland or poolsides), often in rocky places; usually growing in partial shade; 150–950 m. Map 40.

Vouchers: *Compton* 26824; *Faden, Faden & Pooley* 74/209; *Galpin* 1187; *Strey* 8140; *Thorncroft* 9620.

This species has usually been called *A. schlechteri* (below), from which it differs by its longer petioles, larger inflorescences with usually more numerous cincinni, larger flowers, less pubescent pedicels, blue anthers, and dark brown seeds. *Aneilema brunneospermum* similarly differs from *A. arenicola* (above) in its longer petioles, usually more numerously branched inflorescences, larger flowers, less pubescent, usually longer and more recurved pedicels, and dark brown, thicker seeds.

10. **Aneilema schlechteri** *K. Schum.* in Bot. Jb. 33: 376 (1903); Brenan in Kew Bull. 15: 216 (1961), p. p. Type: Transvaal, Komati Poort, 15 December 1897, *Schlechter* 11748 (B, holo.!; BM!; BOL!; BR!; COI!; G!; K!; NSW!; PRE!; S!; Z!, iso.).

Tufted annual. *Roots* thin, fibrous. *Leaves* with leaf-blades sessile to shortly petiolate, lanceolate-elliptic or elliptic to ovate-elliptic or ovate, 15–45 × 10–23 mm, apex acute to obtuse. *Inflorescences* terminal, ovoid thyrses 15–30 × 10–30 mm, with 5–10 ascending cincinni. *Flowers* perfect and staminate; pedicels 2–5 mm long, puberulous almost to base; sepals puberulous, 2–3,2 mm long; petals apparently white; lateral stamen filaments c. 3–4,5 mm long, bearded, anthers yellow; style c. 3–4 mm long. *Capsules* broadly elliptic to obovate-elliptic, bilocular, 3,9–4,5 × 3,4–4,6 mm, greenish tan, locules 1-seeded. *Seeds* elliptic, c. 2,8 × 2 mm, testa light pinkish grey, foveolate-reticulate.

Transvaal; also southeastern Zimbabwe. Basaltic soil in woodland or weed in cultivation (Zimbabwe). Map 41.

Vouchers: *Nel* 5570; *Schlechter* 11748.

This species should be looked for in northern Transvaal. It can be separated from *A. brunneospermum* (above), with which it has been confused, by its sessile or subsessile leaves, smaller inflorescences with generally fewer cincinni, smaller, apparently white flowers on more pubescent pedicels, yellow lateral anthers and lighter coloured seeds.

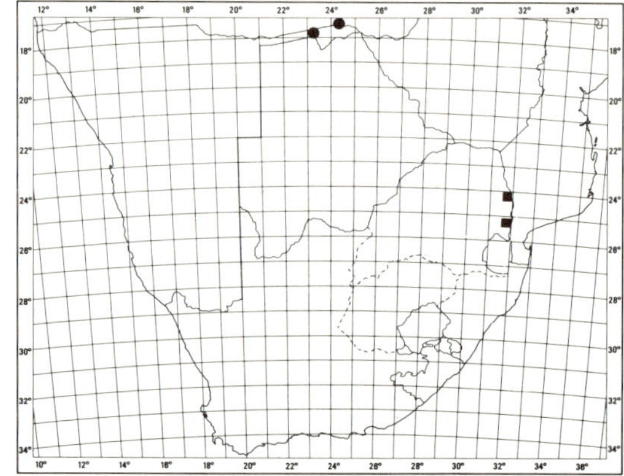

MAP 41.— ● **Aneilema nicholsonii**
　　　　　■ **Aneilema schlechteri**

11. **Aneilema nicholsonii** *C.B. Cl.* in F.T.A. 8: 70 (1901). Type: Zambia(?): road from Missala to Linga Luia River, January, 1897, *Nicholson* s.n. (K!).

Branched annual to 0,6 m tall. *Roots* thin, fibrous. *Leaves* spirally arranged, leaf-blades petiolate, lanceolate-elliptic to ovate-elliptic or ovate, 30–105 × 10–40 mm, apex usually acuminate. *Inflorescences* terminal and axillary from the upper leaves, ovoid to ellipsoid, c. 20–35 × 15–20 mm, dense, usually with c. 20–35 cincinni. *Flowers* perfect and staminate, 8–14,5 mm wide; paired petals blue to blue-purple or lavender, 6–7,5 mm long; lateral stamens densely bearded, 5,5–7 mm long; style 4–6 mm long. *Capsules* oblong-elliptic, trilocular or bilocular, 4,5–6 × 2–3 mm, dorsal locule indehiscent, 1(–0)-seeded, ventral locules usually 3-seeded. *Seeds* of ventral locules ± 3-lobed, 1,25–2,05 × 1,35–1,6 mm, testa tan, foveolate-scrobiculate. Fig. 9: 1.

South West Africa/Namibia (Caprivi Strip); also Zimbabwe and Mozambique to southern Kenya; bushland and woodland, sometimes in rocky places, occasionally in moist situations, often in sandy soil; usually growing in partial shade; c. 1 000 m (in our area). Map 41.

Vouchers: *Curson* 1248; *Venter* 638.

In addition to the bracteole form and densely and conspicuously bearded lateral stamen filaments, *A. nicholsonii* is distinctive because of its trilobed seeds which have a longitudinal groove on the back. The cincinni frequently bear long, uniseriate hairs in addition to shorter, hooked hairs.

899a   **3. MURDANNIA**

Murdannia *Royle*, Illustr. Bot. Himal. 403, t. 95, fig. 3 (1839), nom. cons.; Brenan in Kew Bull. 7: 179 (1952); R. A. Dyer, Gen. 2: 910 (1976). Type species: *M. scapiflora* Royle.

Annual or perennial herbs with sessile leaves; roots various. *Flowers* in terminal or terminal and axillary thyrses or in axillary fascicles, regular or slightly zygomorphic; sepals 3, free; petals 3, free, not clawed, blue, purple or white (rarely yellow). *Stamens* 3–2, fertile, antesepalous; staminodes (4–)3(–0), antepetalous, (when 4, 1 antesepalous). *Ovary* 3-locular, ovules one to many per locule, uniseriate or biseriate. *Capsule* 3-locular; seeds various.

Species c. 50, in S.E. Asia, Australia, S. America and tropical Africa, variable. One species in Southern Africa.

The genus was named after Murdan Aly, a plant collector and keeper of the Herbarium at Saharunpore, whom Royle found to be very knowledgeable on the Himalayan flora.

Murdannia simplex *(Vahl) Brenan* in Kew Bull. 8: 186 "(1952)" 1953; Brenan in J. Linn. Soc., Bot. 59: 349, figs 3 & 9 (1966); Morton in J. Linn. Soc., Bot. 60: 202 (1967); Brenan in F.W.T.A. edn 2, 3: 24, fig. 327 (1968); Schreiber et al. in F.S.W.A. 157: 11 (1969). Faden in Agnew in Upland Kenya Wild Flow. 667, 668 (1974). Type: Guinea, *Thonning* (C, holo.).

Commelina simplex Vahl, Enum. Pl. 2: 177 (1806).

Aneilema sinicum Ker-Gawl. in Bot. Reg. t. 659 (1822), C. B. Cl. in A. DC., Monogr. Phan. 3: 212 (1881), and in F.C. 7: 12 (1897), and in F.T.A. 8: 63 (1901); Hutch. & Dalziel in F.W.T.A. edn 1,2: 312, t. 286 (1936). *Murdannia sinica* (Ker-Gawl.) Brückn. in Pflanzenfam. edn 2, 15a: 173 (1930). Type: ex hortus Kew (seed said to have come from China) (K).

Gregarious, herbaceous, sub-scapose chamaephytes, 0,3–0,6 m tall, from a small hard rhizome, bearing long, hard roots (Note: plants from Natal possess thinner roots). *Leaves* mostly basal, 3–4, linear, up to c. 200 mm long and c. 10 mm broad, smooth, glabrous except for ciliate, open, sheathing basal part. *Flowering shoot* up to 0,5 m tall, the peduncle bearing a few reduced leaves; thyrse terminal or terminal and axillary, with few to many divaricate branches bearing close-set, secund scars of early flowers. *Flowers* slightly zygomorphic, opening in late afternoon for a few hours. *Sepals* 3, ovate, c. 5 mm long, green, persistent. *Petals* 3, equal, obovate, c. 10 mm long, mauve. *Stamens:* 2 upper antesepalous, fertile, and curved downwards, the

3rd sterile; filaments yellow with long, purple, beaded hairs; 3 antepetalous staminodes with glabrous filaments, the empty anthers trilobed. *Ovary* with 2 superposed ovules in each locule; style curved upwards, stigma terminal, small. *Capsule* oblong-globose, c. 6 mm long, shiny; seeds oblong-globose, truncate where they meet, c. 1,5 mm long, the transverse ribs tuberculate, reticulate. Fig. 11.

Recorded from the northern parts of Transvaal, Natal, Botswana and South West Africa/Namibia; widespread in subtropical and tropical Africa to Asia; usually in moist localities. Map 42.

Vouchers: *Killick & Leistner* 3004; *Ross* 2359; *Strey* 4725; *Van der Schijff* 3202, 2036.

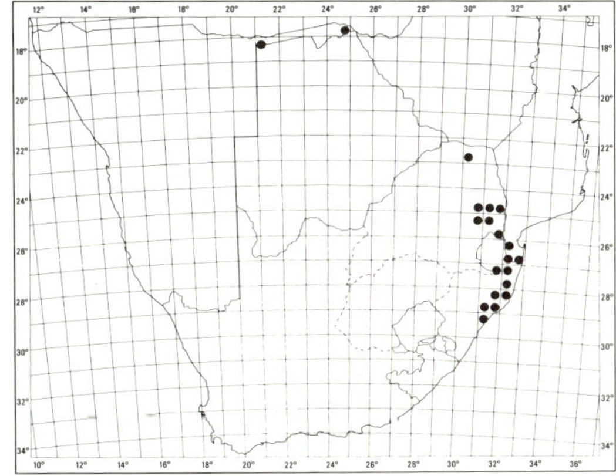

MAP 42.— **Murdannia simplex**

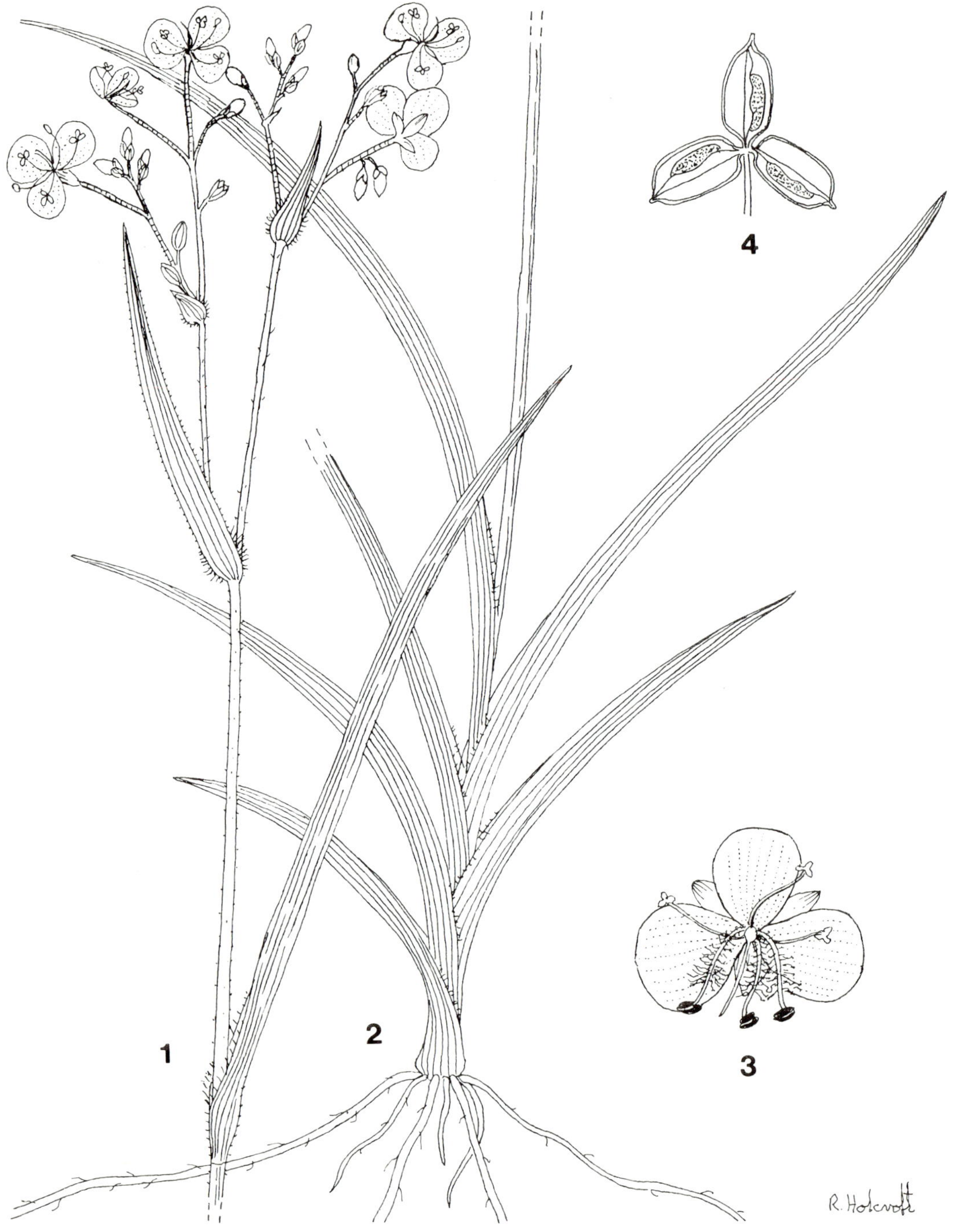

Fig. 11.—**Murdannia simplex**: 1, flowering branch, × 1; 2, basal part of plant, × 1; 3, flower, × 2; 4, open capsule with separated locules, × 2 (*Killick & Leistner* 3004).

FIG. 12.—**Coleotrype natalensis**: 1, flowering shoot, × 1; 2, side view of flower emerging from split spathe below lamina, × 1; 3, flowers from dichasium (after E. Liebenberg in Flower. Pl. Afr.).

4. COLEOTRYPE

903

Coleotrype *C.B. Cl.* in A. DC., Monogr. Phan. 3: 120, 238, t. 8 (1881), and in F.C. 7: 13 (1897); Brückner in Pflanzenfam. edn 2, 15a: 169 (1930); Morton in J. Linn. Soc., Bot. 60: 171 (1967); Perrier in Notul. syst., Paris 5: 196 (1936); Brenan in F.W.T.A. edn 2, 3: 36 (1968); R. A. Dyer, Gen. 2: 911 (1976). Type species: *C. natalensis* C.B. Cl.

Perennial herbs. *Leaves* alternate, sheathing at the base. *Flowers* regular to slightly zygomorphic in axillary, sessile, contracted cymes perforating base of leaf-sheaths. *Sepals* 3, free or fused basally, equal. *Corolla* tubular, limb 3-lobed. *Stamens* 6, equal or unequal, arising from near throat of corolla-tube; filaments usually bearded with long moniliform hairs. *Ovary* 3-locular with 1–2 superposed ovules in each locule. *Capsule* loculicidal, seeds 1–2 per locule, hilum linear, embryotega lateral.

Species 9, in Africa and Madagascar; only 1 in Southern Africa.

The name *Coleotrype*, meaning sheath-borer, refers to the inflorescence piercing through the sheath.

Coleotrype natalensis *C.B. Cl.* in A. DC., Monogr. Phan. 3: 120, 238, t. 8 (1881), and in F.C. 7: 13 (1897); Wood & Evans, Natal Plants 1: 39, t. 48 (1898): Troll in Beitr. Biol. Pflanz. 36: 343 (1961); Oberm. in Flower. Pl. Afr. 37, t. 1465 (1966). Syntypes: Natal, Inanda, *Wood* 479 (K!), *Sanderson* 438 (K!).

Perennials with trailing leafy stems, rooting at the lower nodes, turning upwards, up to 0,8 m tall. *Leaves* alternate, spreading, oblong, c. 100 × 15–20 mm, apex long acuminate, base narrowed into a sheath, margin purple or red. *Inflorescences* in axils of uppermost leaves, sessile, contracted, piercing leaf-sheaths, the 4–6 flowers appearing consecutively in 2 rows. *Sepals* 3, free, resembling hairy bracts, c. 8 mm long. *Corolla* with a narrow tube c. 15 mm long, lobes patent, broadly ovate, c. 16 mm long, deep purple. *Stamens* 6, inserted on corolla-throat, filaments densely bearded above with long, beaded hairs; anthers basifixed with locules curved around discoid connective. *Ovary* slightly hairy above, 3-locular, with 1–2 ovules in each cell; style just exserted, filiform, stigma shallowly cup-shaped with 3 erect, small papillate lobes. *Capsule* obovate-trigonous, 5–7 mm long, rostrate, hairy; seeds oblong-ellipsoid, 4 × 1 mm. Fig. 12.

Recorded from the warmer eastern parts of Southern Africa: Swaziland, Natal, Transkei; also in Mozambique and Zimbabwe; in forested, moist areas often near water, or sandy clearings. Map 43.

Vouchers: *Flanagan* 2503; *Kemp* 596; *Oatley* 28; *Strey* 6821; *Wood* 1564.

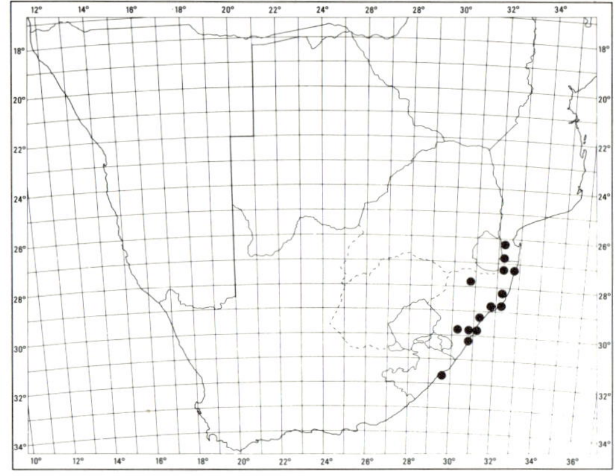

MAP 43.— **Coleotrype natalensis**

904   **5. CYANOTIS**

Cyanotis *D. Don*, Prodr. Fl. Nepal. 45 (1825), nom. cons.; C. B. Cl. in A. DC., Monogr. Phan. 3: 240 (1881); Benth. & Hook.f., Gen. Pl. 3: 851 (1883); C. B. Cl. in F.C. 7: 13 (1897), and in F.T.A. 6: 78 (1901); R. A. Dyer, Gen. 2: 911 (1976). Type species: *C. barbata* D. Don.

Succulent perennials or (rarely) annuals, the perennials commonly with storage organs such as bulbs, corms, rhizomes or tubers and often with separate fertile and sterile shoots. *Roots* numerous, thin or stout, sometimes tuberous. *Leaves* succulent, spirally arranged or distichous, sessile. *Inflorescences* consisting of terminal or terminal and axillary cymes, each cyme subtended by a spathaceous, leaf-like bract and consisting of two-ranked, sessile or subsessile flowers and usually conspicuous, herbaceous bracteoles. *Flowers* regular, bisexual, small, open for a few hours only, in the morning. *Calyx* tubular below, with somewhat hairy lobes, persistent. *Corolla* with a short tube, the 3 lobes erect or spreading, acute to obtuse, blue, purple or pink. *Stamens* 6, equal, erect, white or coloured like corolla or darker; filaments filiform, or sometimes fusiform in upper half, long, patent, beaded hairs always present in upper half; anthers yellow, basifixed, locules opening from a basal aperture, then splitting open to apex.

Ovary ovoid, 3-locular, hairy above; ovules 2 in each locule, superposed; style thin, filiform or in some species fusiform below small apical stigma, glabrous or bearded. *Capsule* narrowly ovoid, erect; seeds oblong-globose, obtuse, embryotega at apex of upper seed, and at base of lower, testa greyish brown, wrinkled or pitted, epidermis minutely pellucid-dotted.

Species about 45, warm regions of Africa, Asia and northern Australia; 7 in Southern Africa; absent from winter rainfall region.

C. B. Clarke in F.C. 7: 14 described the petals as free at the base and then connate into a tube; in the specimens studied the corolla-tube was fused from the base.

The name *Cyanotis* refers to the blue flowers found in the genus.

1 Plants forming basal leaf-clusters, rhizomatous or with a hard compact root-crown:
  2 Fertile shoots decumbent, usually with many flower-clusters; forming colonies, saxicolous:
    3 Plants with thin roots from creeping rhizomes .... 1. *C. lapidosa*
    3 Plants with thick roots:
      4 Roots emerging from a vertical rhizome; leaves broadly linear, up to 20 mm broad; cymes sessile .... 2. *C. robusta*
      4 Roots emerging from a compact root-crown; leaves narrowly linear, c. 3 mm broad; cymes often pedunculate .... 3. *C. pachyrrhiza*
  2 Fertile shoots erect, usually with 4–6 flower-clusters widely spaced; plants solitary or few together; grassland:
    5 Rhizome vertical to U-shaped; roots terete, hard; pubescence of short patent hairs .... 4. *C. speciosa*
    5 Rhizome a small hard root-crown; roots spindle-shaped, swollen; pubescence of soft appressed hairs ....
    .... 5. *C. longifolia*
1 Plants without a basal leaf-cluster and without a rhizome or root-crown:
  6 Leaves not clasping the stem; spathes usually 2 at some or all nodes; cymes borne at 1–3(–4) nodes on shoot; annuals .... 6 *C. lanata*
  6 Leaves clasping the stem; spathes 1 at each node; cymes borne at 5–15 nodes on shoot; perennials   7. *C. foecunda*

1. **Cyanotis lapidosa** *Phill.* in Flower. Pl. Afr. t. 318 (1928). Type: Transvaal, Kaapse Hoop, *Phillips* 3449 (PRE, holo.!).

Small, spreading rhizomatous perennials often with light to dark vinaceous colouring at base, saxicolous, forming colonies; roots thin, branched. *Sterile shoots* perennial, with limited growth, forming a basal cluster of c. 7 subdecussate to rosulate leaves; lamina sheathing basally, broadly linear, c. 150 × 12 mm, upper smaller (immature), lanate with long, white, appressed hairs, rarely glabrous; further apical growth suppressed towards time of flowering. *Fertile shoots*, annual, arising below basal leaf-cluster, forming long, decumbent, spreading stems with the narrowly ovate, leaf-like spathes diminishing in size towards apex, and enclosing compact, few–many-flowered sessile cymes in their axils; lower nodes producing 1–2 short adventitious side branches which pierce lower empty spathes on adaxial side; roots, if developed, similarly adventitious. After flowering (usually) new decumbent shoots sprout from adventitious buds below leaf-cluster; these give rise to new clusters, the plants thus in time forming large colonies. *Flowers* typical; corolla c. 10 mm long, purple, mauve or pink.

Stamens with filaments fusiform towards apex, where they are densely covered with beaded hairs, darker in colour than petals. *Capsule* with seeds typical, 1,5 mm long. Fig. 13.

Widespread and common in Transvaal, Swaziland and northern Natal; on rocky ledges, forming colonies. Map 44.

Vouchers: *Bredenkamp* 803; *Buitendag* 455; *Codd* 8052; *Compton* 25603; *Mauve* 5269; *Repton* 492.

Unfortunately collectors often omit to gather the whole plant, picking only the annual flowering shoots.

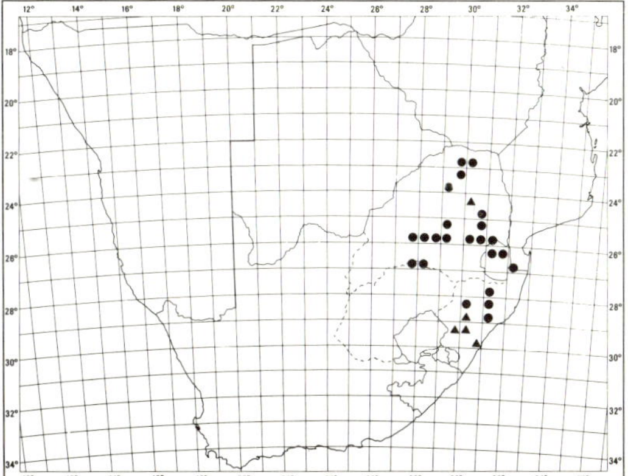

MAP 44.— ● Cyanotis lapidosa
▲ Cyanotis robusta

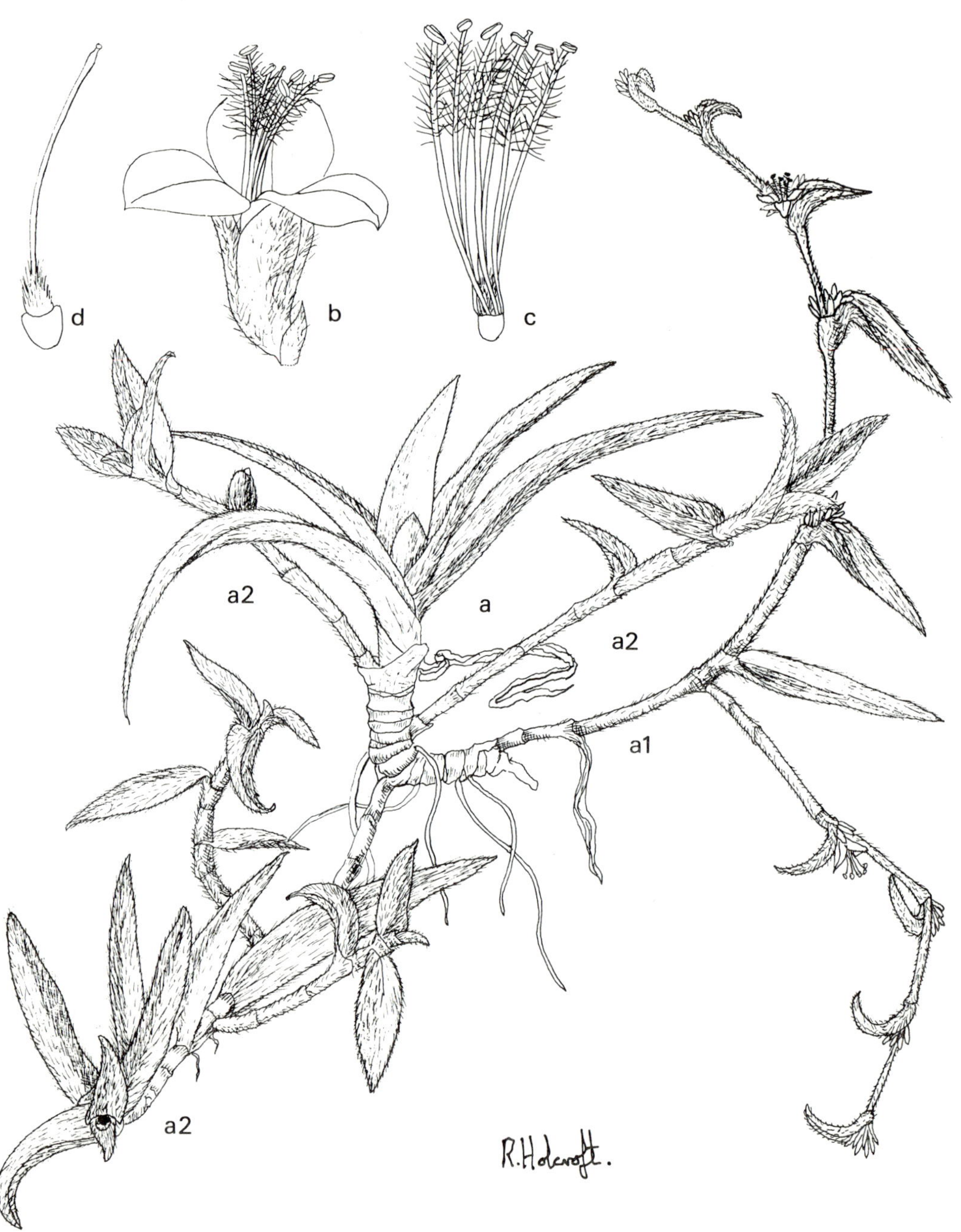

FIG. 13.—**Cyanotis lapidosa**: a, sterile shoot with leaf-cluster; a1, annual flowering shoot with side branch emerging from a tear in a leaf-sheath, × 0,8; a2, vegetative shoot, × 0,8; b, flower, × 3; c, stamens, × 3; gynoecium, × 3 (*Mauve* 5284).

2. **Cyanotis robusta** *Oberm.* in Bothalia 13: 438 (1981). Type: Natal, near Weenen, *Arnold* 1372 (PRE, holo.!).

Plants robust, rhizomatous, lanate, forming colonies, saxicolous. *Rhizome* short and thick, vertical, bearing long, thick roots, placed close together at the base, c. 5 mm in diam., in turn producing active, thin, branched rootlets. *Leaves* in basal clusters, sub-rosulate, c. 6, linear-acuminate, c. 120–300 × 13–26 mm, flat, somewhat fleshy, glabrous and shiny above, lanate below, with long, thin, appressed, white hairs. *Fertile shoots* arising below (sterile) leaf-cluster, procumbent, many-noded, upper bearing dense, lanate cymes enclosed by ovate, falcate spathes; bracteoles small. *Flowers* typical. *Calyx* with erect hairy lobes, c. 10 mm long. *Corolla* 15(–20?) mm long, blue. *Capsule* narrow with an apical tuft of setae; seeds oblong-globose, c. 2 mm long, typical.

Two records from the eastern Transvaal (Wolkberg); otherwise recorded from Natal, in the vicinity of Estcourt and Weenen; on rocky ledges of sandstone krantzes. Map 44.

Vouchers: *Mogg* 7229; *Müller & Scheepers* 165; *West* 1441, 1538.

3. **Cyanotis pachyrrhiza** *Oberm.* in Bothalia 13: 437 (1981). Type: Transvaal, Ohrigstad Dam, *Mauve & Retief* 5245 (PRE, holo.!).

Perennials forming mats; with purple colouring and soft patent white pubescence on leaves and flowering stems. *Roots* long, thick, c. 3 mm in diam., tapered below, white (dark when dried), emerging from a small compact root-crown. *Leaf-cluster* with 5–7 erect linear leaves up to 150 × 8 mm. *Fertile shoots* 4–5-noded, erect or spreading, up to c. 200 mm long, bearing few- to many-flowered axillary cymes, pedunculate or sessile; subtending spathe long and narrow, recurved, not enveloping cyme, occasionally spathe short; bracteoles short. *Calyx* with lobes fused at the base, c. 7 mm long. *Corolla* just exserted from calyx, maroon ("petunia violet": *Ridgway*), lobes semi-erect, triangular. *Stamens* with maroon, beaded hairs, filaments white, fusiform near apex, anthers yellow. *Style* fusiform below stigma. *Capsule* subquadrate, sparsely hairy in upper half; seeds oblong-globose, c. 1,5 mm long, wrinkled.

So far only recorded from the eastern Transvaal, with one record from the N.W. Transvaal; in montane areas, on quartzite ledges. Flowering December. Map 45.

Vouchers: *Breyer* sub TRV 17813; *Codd* 8052; *Retief & Herman* 106; *Smuts & Gillett* 2434; *Van Dam* sub TRV 26303.

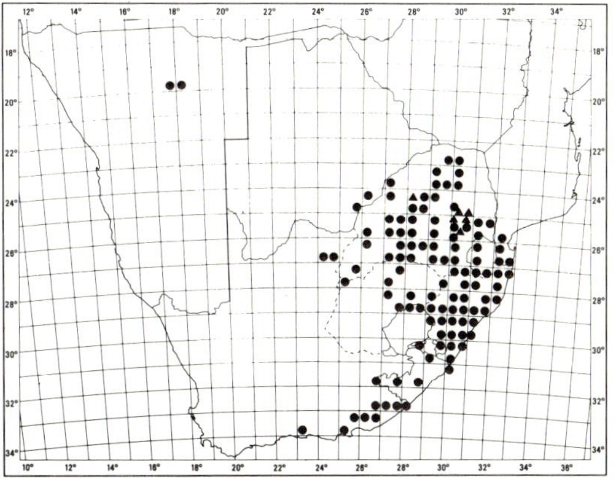

MAP 45.— ● **Cyanotis speciosa**
      ▲ **Cyanotis pachyrrhiza**

4. **Cyanotis speciosa** *(L.f.) Hassk.* in Commel. Ind. 108 (1870). Type: C.B.S., *Thunberg* (LINN 406.8, holo., PRE, photo.!).

Tradescantia speciosa L.f., Suppl. 192 (1781). *Commelina speciosa* (L.f.) Thunb., Prodr. 58 (1794), and Fl. Cap. edn 2: 294 (1823).

Cyanotis nodiflora (Lam.) Kunth, Enum. 4: 106 (1843); Hook. in Curtis's bot. Mag. 20, t. 5471 (1864); C. B. Cl. in F.C. 7: 14 (1897), and in F.T.A. 8: 82 (1901); Brückner in Pflanzenfam. edn 2, 15a: 167, f. 62A (1930); Schreiber in F.S.W.A. 157: 10 (1969). *Tradescantia nodiflora* Lam., Encycl. 2: 371 (1786). Type: C.B.S., *Sonnerat* (P, holo.).

Sonnerat's specimen was probably collected in the eastern Cape. Names: Khunyela (Zulu); damba (Venda).

Perennials (chamaephytes in colder areas), usually solitary plants or a few together, variable in size, with the basal vegetative shoot bearing c. 5 sub-decussate leaves and erect flowering-stems c. 300–500 mm tall. *Rhizome* well developed, swollen, perpendicular (in soft soil), forming a U-shaped body, the older part dying, the new erect shoot sprouting from base (if in shallow or hard soil rhizome will be horizontal). All shoots enclosed in tight sheaths, at first purple-coloured. *Leaf-bearing shoot* forming a cluster of c. 5 sub-decussate leaves, linear-acuminate, c. 170 × 8 mm, canaliculate, glabrous above, with long patent hairs below.

FIG. 14.—**Cyanotis speciosa**: 1 & 1a, habit, showing geniculate rhizome and flowering branch; a–d fertile shoots (leaf-sheaths removed); e, bud of new rhizome, all × 0,6; 2, base of younger plant with leaf-sheaths in place, × 0,6; 3, flower, × 4; 4, stamen, × 4; 5, gynoecium, × 4; 6, beaded hair of filament, much enlarged (all from *Mauve & Holcroft* 5244).

Fertile shoots 1–3, erect, firm, terete, with 1–2 sterile bracts below, above with falcate spathes, subtending c. 3–4 dense, axillary cymes. *Flowers* pale to deep oriental blue *(Ridgway)* or mauve, c. 10 mm long. *Capsule* obovoid, c. 5 mm long, hairy at apex; seeds oblong-globose, 2 mm long, light brown, shiny, wrinkled. Fig. 14.

Widespread and often common locally as solitary plants or few together; Southern Africa (summer rainfall region) to southern Tanzania and Madagascar. Growing in grasslands; flowering intermittently November–December. "Tubers relished by pigs" (*Galpin* 371). Map 45.

Vouchers: *De Wet* 1701; *Harbor* sub TRV 19201; *Liebenberg* 8389; *Pegler* 1149B; *Ward* 6466.

The species *Cyanotis gryphaea* Dinter in Reprium nov. Sp. Regni Veg. 16: 365 (1920), nom. subnud., may be a synonym. The type is *Dinter* 2424 from South West Africa/Namibia, Gaub (not seen).

5. **Cyanotis longifolia** *Benth.* in Hook., Niger Fl. 543 (1849); C. B. Cl. in A. DC., Monogr. Phan. 3: 259, excl. var. *caespitosa* C. B. Cl. (1881), and in F.T.A. 8: 81 (1901); Brenan in Kew Bull. 1952: 205 (1953) and in F.W.T.A. edn 2,3: 37 (1968); Morton in J. Linn. Soc., Bot. 60: 195 (1967); Schreiber et al. in F.S.W.A. 159: 10 (1969). Type: Congo, *Curror* 1 (K, holo.!).

Plants c. 350(–900) mm tall, erect with a lanate pubescence of long, thin, straight, soft hairs, especially dense at base of leaf-cluster. *Roots* fusiform, long, swollen part, when young, densely covered by root hairs. *Leaf-cluster* with c. 5 long, erect, flat, linear leaves, c. 300 × 10 mm. *Fertile shoots* c. 450 mm tall with few stem-leaves reduced above to spathes bearing a few terminal and axillary, dense, sickle-shaped cymes; spathes recurved. *Flowers* with calyx-lobes bearing an apical tuft of long erect hairs or lanate all over; corolla blue; filaments with blue, beaded hairs, fusiform apically; style bearded. *Capsule* oblong-globose, c. 5 mm long; seeds typical.

A variable species recorded from northern South West Africa/Namibia; common in tropical Africa but apparently rare in the south. Map 46.

Vouchers: *De Winter* 3939; *De Winter & Wiss* 4088.

6. **Cyanotis lanata** *Benth.* in Hook., Niger Fl. 542 (1849); C. B. Cl. in A. DC., Monogr. Phan. 3: 258, includ. both vars (1881), and in F.T.A. 8: 80 (1901); Morton in J. Linn. Soc., Bot. 60: 194 (1967); Brenan in F.W.T.A. edn 2, 3: 40 (1968). Syntypes: Nigeria, Lower

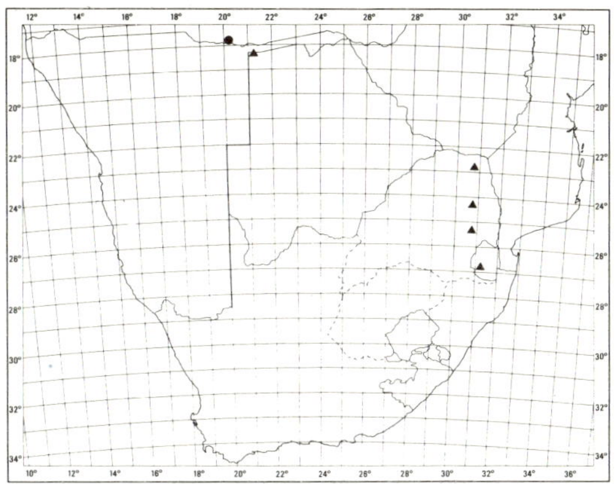

MAP 46.— ● **Cyanotis longifolia**
      ▲ **Cyanotis lanata**

Niger: Patteh Mountain, *Vogel* 183 (K); Quorra (River Niger), *Vogel* 122 (K!).

Annuals, erect and few-branched at first, later densely branched and covered with long, soft, appressed hairs, rarely glabrous. *Leaves* narrowly oblong, acute, 30–60 × 4 mm, narrowed into a tubular sheath. *Cymes* few-flowered, clustered in axils of spathes. *Corolla* small, violet, blue, pink or mauve; filaments nearly glabrous or with blue hairs, anthers orange-yellow. *Capsule* triangular, obovoid, thin, 5 mm high; seeds oblong, 1–2 mm long, punctate with raised areas.

Widely distributed in the warmer parts of Africa. Recorded from eastern Transvaal and Swaziland; in rock fissures or in cultivated lands. Map 46.

Vouchers: *Buitendag* 456; *Karsten* sub PRE 38431; *Strey* 3620.

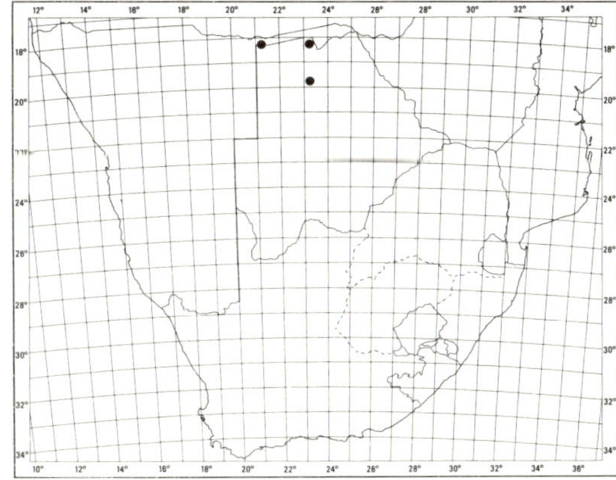

MAP 47.— **Cyanotis foecunda**

7. Cyanotis foecunda *Hassk.*, Commel. Ind. 110 (1870). C. B. Cl. in A. DC., Monogr. Phan. 3: 255 (1881), and in F.T.A. 8: 80 (1901). Type: Ethiopia, Seraba in Uschan, *Schimper* 459 (K, iso.!).

Soft, hirsute herbs with weak spreading annual stems forming long internodes, rooting at nodes. A small bulb, apparently deep-seated in ground and easily detached, may be present at base. *Leaves* broadly linear, c. 80 × 10 mm, apex acute, abruptly clasping stem at base; spathes progressively smaller towards shoot apex, ovate, bearing sessile, few-flowered cymes. *Corolla* blue or violet. *Stamens* with filaments bearded with blue hairs and fusiform below anthers. *Style* usually glabrous. Capsule not seen.

Recorded from northern South West Africa/Namibia and northern Botswana; common in Zimbabwe to tropical East Africa and Ethiopia; in grasslands, on termite hills or in rock crevices, said to form dense mats on occasion. Map 47.

Vouchers: *De Winter & Marais* 4828; *Smith* 4090, 1629.

6. FLOSCOPA

908

Floscopa *Lour.*, Fl. Cochin. 189, 192 (1790); C. B. Cl. in A. DC., Monogr. Phan. 3: 267 (1881), and in F.C. 7: 14 (1897), and in F.T.A. 8: 84 (1901); Brenan in F.W.T.A. edn 2,3: 26 (1968); R. A. Dyer, Gen. 2: 911 (1976). Type species: *F. scandens* Lour.

Perennial or annual hygrophytic herbs. *Stems* erect or ascending, rooting from lower nodes or with a basal root-crown. *Leaves* mostly linear to oblong, acute, sheathing at base. *Inflorescence* a terminal, lax to dense, often leafy compound thyrse, glandular-pubescent or glabrous; bracteoles small or lacking; pedicels short. *Flowers* small, bisexual. *Sepals* 3, free, equal, persistent. *Petals* 3, free, deliquescent, equal or lower one narrower, blue, mauve, pink or yellow. *Stamens* (5–)6, equal or upper 3 slightly different from lower 3; filaments glabrous; upper 3 fused basally. *Ovary* 2-celled with 1 ovule in each locule, glabrous. *Capsules* shortly stipitate, dorsiventrally compressed, ellipsoid to obovoid, apex retuse or acute; seeds depressed hemispherical, ribbed or smooth, occasionally covered by small glandular discs, with a linear hilum and dorsal embryotega.

Widespread in warmer regions of Africa, Asia, Australia, Central and South America; c. 20 species. Three species in Southern Africa; aquatics.

The name was derived from *flos* (flower) and *scopa* (broom), as the compact inflorescence resembles a flower-broom.

1 Flowers purple to pink; moderate-sized, sprawling plants 0,3–1 m high with cauline leaves:
  2 Inflorescences glandular-pubescent, usually dense .... 1. *F. glomerata*
  2 Inflorescences glabrous, lax .... 2. *F. leiothyrsa*
1 Flowers yellow to orange; small annuals up to c. 100 mm high with basal leaves .... 3. *F. flavida*

1. Floscopa glomerata *(Willd. ex J.A. & J.H. Schult.) Hassk.* in Commel. Ind. 166 (1870); C. B. Cl. in A. DC., Monogr. Phan. 3: 267 (1881), and in F.C. 7: 15 (1897), and in F.T.A. 8: 86 (1902); Brenan in F.W.T.A. edn 2,3: 28 (1968); Schreiber et al. in F.S.W.A. 159: 11 (1969). Type: Madagascar, collector ? (B, herb. Willdenow, microfiche 6345).

Tradescantia glomerata Willd. ex J. A. & J. H. Schult., Syst. Veg. 7(2): 1175 (1830). *Dithyrocarpus glomeratus* (J.A. & J.H. Schult.) Kunth, Enum. Plant. 4: 78 (1843).

D. capensis Kunth, Enum. Plant. 4: 78 (1843). Type: Cape, *Drège*, Herb. Cap. No. 4472 (B, holo.).

Perennial aquatic herbs exserted to c. 0,6 m above water level. *Stems* few-branched, decumbent, producing numerous adventitious roots from each lower node. *Leaves* linear-acuminate, c. 80 × 12 mm, basal cylindrical sheath c. 12 mm long; upper younger leaves progressively smaller, glabrous. *Inflorescences* terminal, glandular-pubescent, compact, c. 20–40 mm long, consisting of several to many one-sided, dense, short cymes and reduced thyrses, with flowers close together, ebracteolate, shortly pedicelled. *Flowers* small, zygomorphic, open in morning for short period. *Sepals* ovate, 2–3 mm long, pale purple, glandular-pubescent. *Petals* 3, mauve, unequal, upper 2 ovate, c. 3–4 mm long, lower linear. *Stamens* 6; 3 lower longer, declinate, anthers with narrow connectives; 3 upper somewhat shorter, anther sacs round, separated by a broad, square, yellow connective. *Ovary* compressed-cordate; style long, filiform; stigma terminal, minute, included within shrivelled petals after anthesis. *Capsule* c. 2 mm long, pale, shiny; seed hemi-ovoid, 1–1,5 mm long, smooth, greyish blue. Fig. 15.

Widespread in warmer parts of Southern Africa to tropical Africa and Madagascar; growing along streambanks. Map 48.

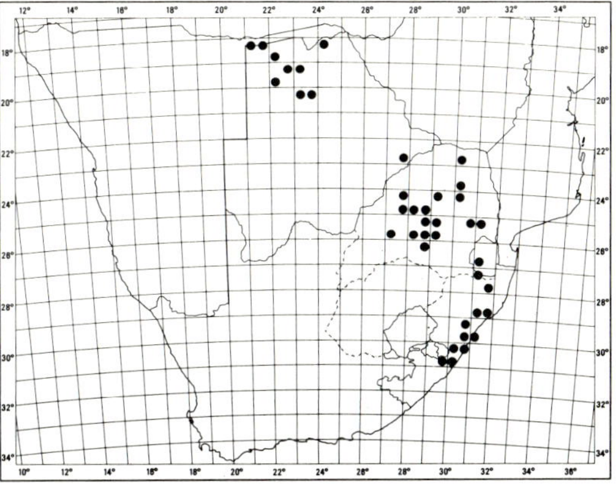

MAP 48.— **Floscopa glomerata**

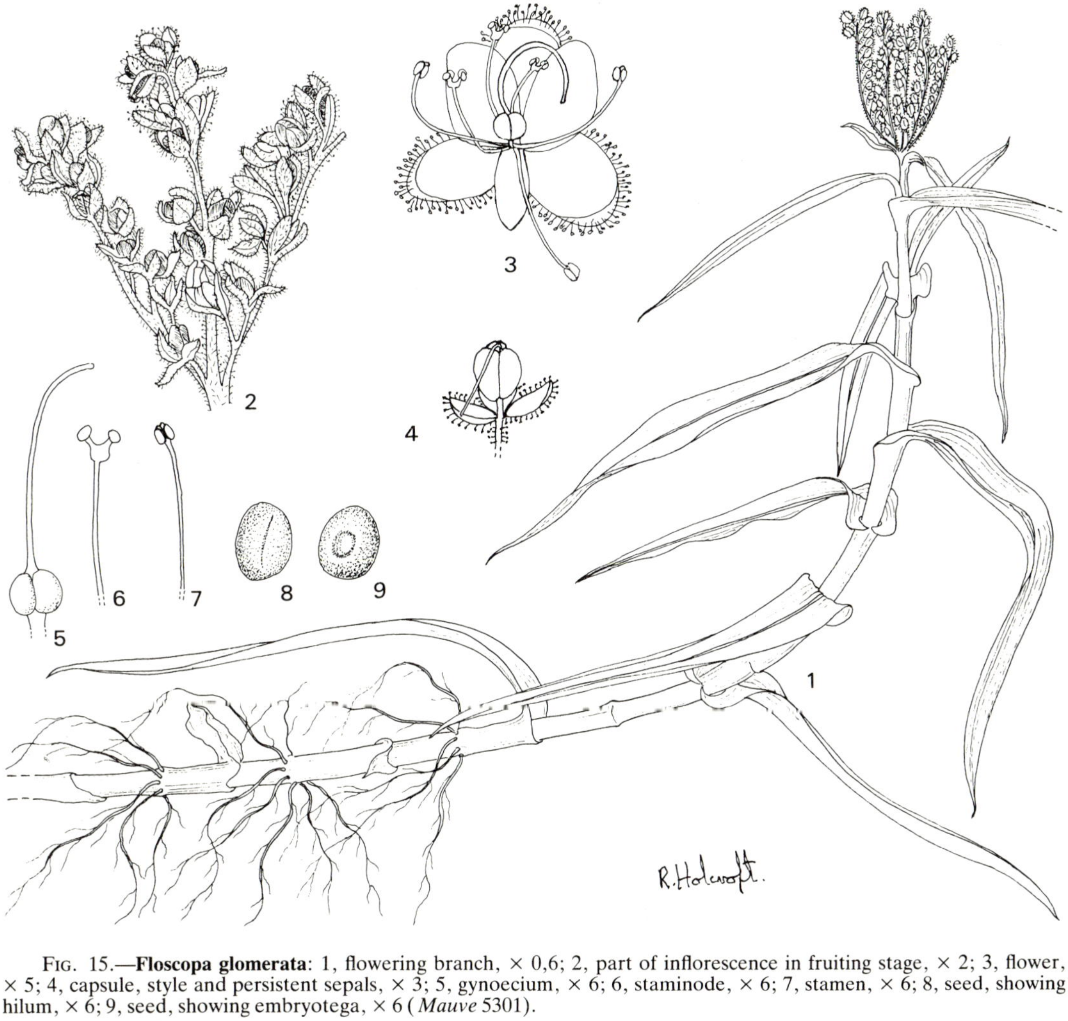

FIG. 15.—**Floscopa glomerata**: 1, flowering branch, × 0,6; 2, part of inflorescence in fruiting stage, × 2; 3, flower, × 5; 4, capsule, style and persistent sepals, × 3; 5, gynoecium, × 6; 6, staminode, × 6; 7, stamen, × 6; 8, seed, showing hilum, × 6; 9, seed, showing embryotega, × 6 (*Mauve* 5301).

Vouchers: *Acocks* 13371; *Junod* sub TRV 21220; *Merxmüller & Giess* 1953; *Story* 4793; *Strey* 3270; *Wood* 6372.

2. Floscopa leiothyrsa *Brenan* in Kew Bull. 7: 206 (1952), and in F.W.T.A. edn 2, 3: 27 (1968). Type: Tanzania, Dodoma district, Chaya Lake, *Burtt* 3802 (K, holo.).

Annual, weak-stemmed glabrous herb, somewhat succulent, up to 600 mm tall with sparse, thin, short, spreading side branches. *Leaves* few, sessile, narrowly linear-acuminate, c. 70 × 4 mm, upper changing to bracts. *Inflorescence* of loose cymes, in axillary and terminal thyrses. *Sepals* 3, convex, elliptic. *Petals* blue, purple or carmine, upper 2 obovate-elliptic, 3 mm long, somewhat unequal, lower one oblong-linear, 3,5 mm. *Stamens* 3 + 3, unequal. *Ovary* compressed. *Capsule* narrowly obovate, 2 mm long, pale; seeds hemi-ellipsoid, 1 mm long, black, shiny, finely reticulate.

Recorded from tropical Africa to northern Botswana; in running water. Map. 49.

Vouchers: *Smith* 1778; *Story* 4760.

3. Floscopa flavida *C.B. Cl.* in A. DC., Monogr. Phan. 3: 269 (1881), and in F.T.A. 8: 87 (1901); Brenan in F.W.T.A. edn 2, 3: 27, f. 328 (1968). Syntypes: Central Africa, Djur, *Schweinfurth* 2537, 4286; Dahomey, Borgu near Niger River, *Barter* 760.

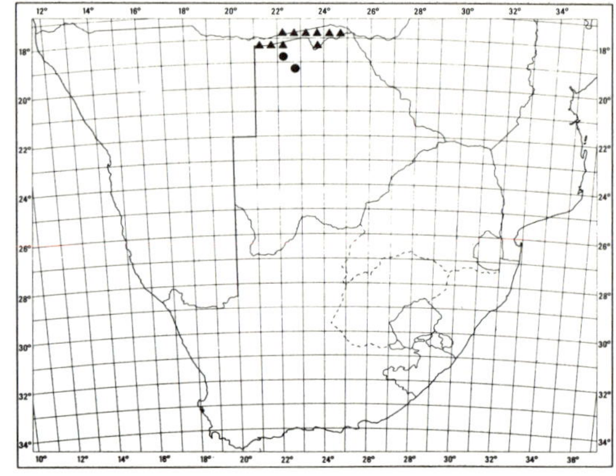

MAP 49.— ● **Floscopa leiothyrsa**
▲ **Floscopa flavida**

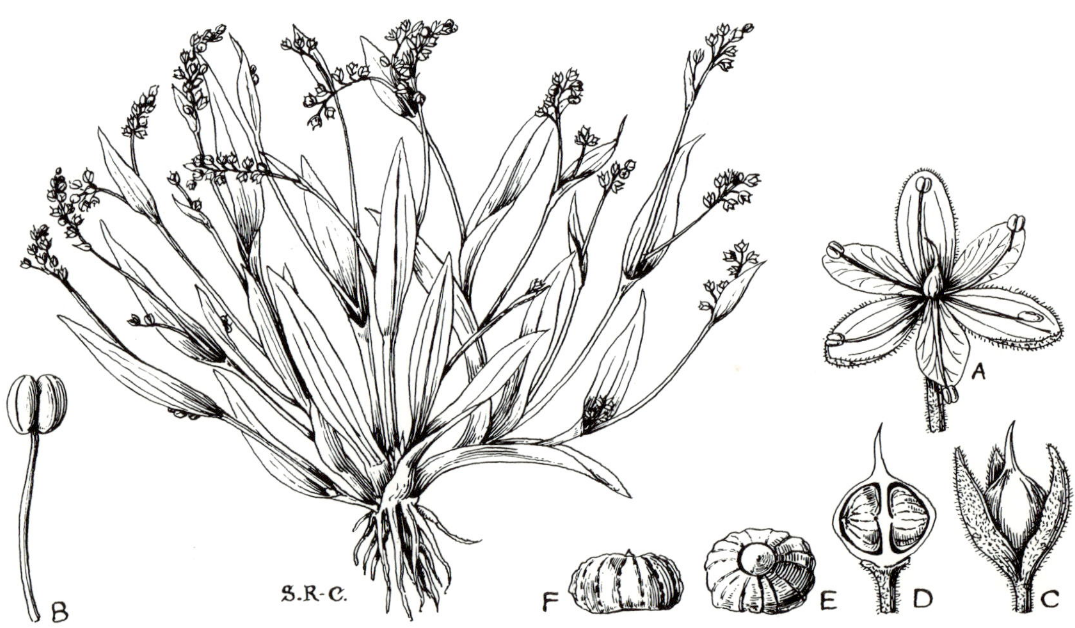

FIG. 16.—**Floscopa flavida**: A, open flower, × 3; B, stamen, × 8; C, fruit, × 8; D, vertical section of fruit, × 8; E, top and F, side view of seeds, × 10. Reproduced from the 'Flora of West Tropical Africa' with permission of the Director of the Royal Botanic Gardens, Kew.

Annuals, c. 100 mm tall, with basal, erect, light green leaves and terminal and axillary thyrses. *Roots* many, long, yellow, with numerous thin side roots. *Leaves* linear, acute, c. 50–80 × 7 mm, glabrous, loosely sheathing at the base. *Inflorescences* on thin exserted peduncles, composed of crook-shaped cymes or flowers more loosely arranged; small, broad bracteoles sometimes present. *Sepals* 3, maroon to purplish, ovate, 2 mm long. *Petals* yellow to orange, equal, slightly longer than the sepals. *Stamens* 6, equal. *Ovary* typical. *Capsule* compressed, broader than long, 2 mm broad, apiculate, shiny, cream; seeds subglobose, ribbed, reticulate and punctulate, cream to grey. Fig. 16.

Widespread in warmer parts of Africa; Botswana, South West Africa/Namibia (Caprivi Strip); in marshy areas. Map 49.

Voucher: *Curson* 13.

911

7. TRADESCANTIA

Tradescantia *L.*, Sp. Plant, 288 (1753); Brückner in Pflanzenfam. edn 2, 15a: 166 (1930); Bailey, Stand. Cycl. Hort. edn 20: 3363 (1963). Type species: *T. virginiana* L.

Perennial herbs. *Stems* simple to diffusely branched, erect or trailing, sometimes rooting at nodes. *Leaves* oblong-ovate to linear. *Inflorescence* terminal and/or axillary, composed of paired, sessile cymes, each pair subtended by leaf-like or spathe-like bracts. *Flowers* regular, pedicellate, few to numerous. *Sepals* 3, free, green or coloured. *Petals* 3, free or connate, obovate to orbicular, sometimes clawed, blue, rose, purple or white. *Stamens* 6, equal, filaments bearded or smooth. *Ovary* 3-celled, with 2(–1) superposed ovules in each cell. *Capsule* loculicidally dehiscent; seeds variable, hilum linear to punctate; embryotega dorsal.

About 60 species in North and South America; several in cultivation. One or two species naturalized in Southern Africa, as *T. virginiana* L. has also been recently recorded as a garden escape. The genus was named after John Tradescant, gardener to Charles I, who died in 1638.

Tradescantia fluminensis *Vell.*, Fl. Flum. 140, 3, t. 152 (1827); Bailey, Stand. Cycl. Hort. edn 20: 3363, t. 3829, 3830 (1963); Hortus Third 1120 (1977); Cheesman, Man. New Zeal. Fl. 1059 (1925); Adamson in Adamson & Salter, Fl. Cape Penins. 160 (1950). Iconotype: Brazil, t. 152 in Fl. Flum.

Glabrescent herbs with decumbent, slender, leafy stems, rooting at nodes. *Leaves* distichous, ovate-acuminate, abruptly narrowed into a short, broad, open, ciliate sheath, c. 25–40 × 15–20 mm, often with white and purple stripes, purple below. *Flowers* in pairs of few-flowered cymes, terminal and/or terminating abbreviated side branches; each pair of cymes subtended by 2 leaf-like subequal bracts, pedicels c. 10 mm. *Sepals* ovate-acuminate, with a ciliate keel, c. 6 mm long, green. *Petals* ovate, c. 8 mm long, white. *Stamens* 6, with filaments c. 8 mm long, bearing long beaded hairs in lower half; anthers with an obtriangular connective, locules spreading outwards towards apex. *Ovary* 3-locular, oblong-globose, with 2 ovules in each locule; style terete or somewhat swollen in middle, stigma capitate. *Capsule* 2 mm long, chartaceous; seeds reticulate, hilum linear.

South America: Central Brazil to Argentina. A common weed in New Zealand. So far rare in Southern Africa: a few records from Natal, Swaziland and eastern Transvaal; shade-loving. One of several species often cultivated and referred to as "Wandering Jew". Map 50.

Vouchers: *Buitendag* 1120; *Miller* 3040; *Ram* s.n.

Often mistaken for *T. albiflora* Kunth, a species commonly cultivated in Southern Africa, which is coarser and has no purple colouring on the lower side of the leaf. However, they may represent different forms of a single species, as suggested by Dr R. B. Faden.

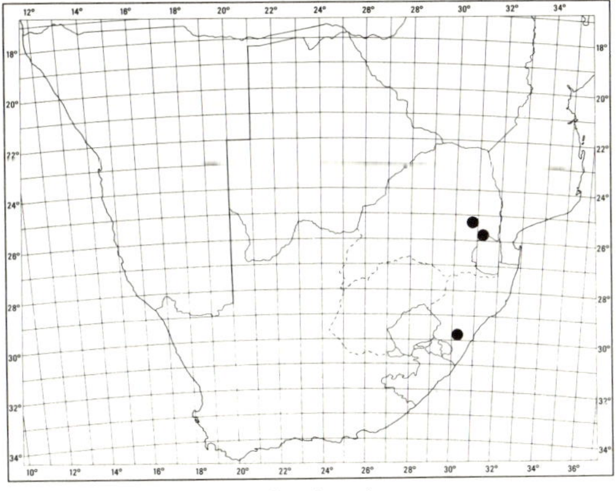

MAP 50.—**Tradescantia fluminensis**

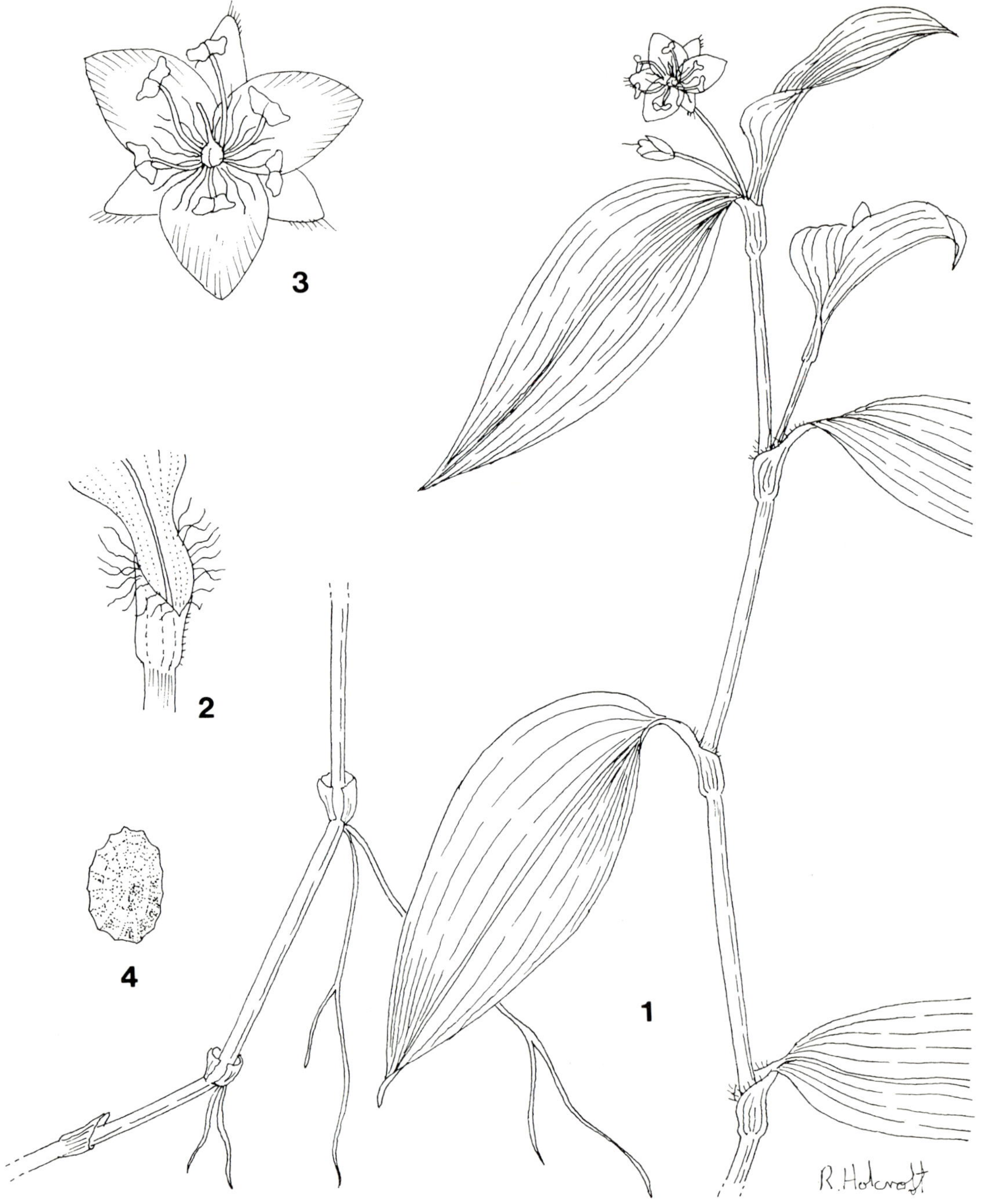

FIG. 17.—**Tradescantia fluminensis**: 1, base of plant and flowering branch, × 1; 2, leaf-sheath, × 1,5; 3, flower, × 5; 4, seed, × 10 (*Buitendag* 1120).

PONTEDERIACEAE

by A. A. OBERMEYER

Aquatic herbs with sympodial stems or rhizomes, rooted or free-floating. *Leaves* often dimorphous, spirally arranged, sheathing at base; emergent leaves with long, sometimes swollen petioles; blades spathulate, orbicular, ovate, cordate or hastate; submersed leaves linear. *Inflorescence* terminating a 1-leaved section of sympodium, exserted from leaf-sheath, racemose, spicate, subumbellate or 2–1-flowered. *Flowers* ebracteate, bisexual, occasionally some cleistogamous. *Perianth* usually with a short tube and spreading zygomorphic limb, rarely 6-partite and more or less regular, blue, white or yellow, fugaceous, marcescent. *Stamens* (1) 3–6, dimorphous, filaments free, inserted on perianth-tube at different levels or hypogynous, anthers 2-thecous, opening by slits or pores. *Ovary* superior, (1)3-locular with axile or parietal placentas, 1–many-ovulate; style filiform; stigma capitate or shortly lobed. *Capsule* loculicidal, 1–3-valved, with a thin pericarp or a utricle; seeds small, ovate, obtuse, ribbed, spaces between ribs with close-set transverse lines.

Genera 5, species about 25, all freshwater hygrophytes, widespread in tropics and subtropics.

Lit. Solms-Laub. in A. DC., Monogr. Phan. 4: 501 (1883), and in Bot. Ztg 41: 301 (1883); Benth. & Hook. f., Gen. Pl. 3: 836 (1883); Schonl. in Natürl. PflFam. edn 2,4: 70 (1888); N.E. Br. in F.C. 7:1 (1897), and in F.T.A. 8: 1 (1902); Schwartz in Beih. bot. Zbl. 42,1: 263 (1926), and in Bot. Jb. 61, Beibl. 139: 28 (1927), and in Natürl. PflFam. edn 2, 15a: 181 (1930); Sculthorpe, Biol. Aquat. Vasc. Plants (1967); Hepper in F.W.T.A. edn 2,3,1: 108 (1968); Verdc. in F.T.E.A. Pontederiaceae: 1 (1968); Podlech in F.S.W.A. 154: 2 (1969); Cook et al., Waterplants World 482 (1974); R. A. Dyer, Gen. 2: 912 (1976).

1 Ovary 1-locular, with one ovule; utricle enclosed in accrescent base of perianth   3. **Pontederia**
1 Ovary 3-locular, with numerous ovules; capsule many-seeded:
  2 Stamens 3 .... 4. **Heteranthera**
  2 Stamens 6:
    3 Perianth-segments free to base; one stamen larger with a blue anther and an erect apical tooth .... 1. **Monochoria**
    3 Perianth forming a tube and spreading zygomorphic limb; with 3 short and 3 long stamens .... 2. **Eichhornia**

920

1. MONOCHORIA

Monochoria *Presl*, Reliq. Haenk. 1: 127 (1827); N.E. Br. in F.T.A. 8: 51 (1901); Hepper in F.W.T.A. edn 2, 3,1: 108 (1968); Verdc. in F.T.E.A. Pontederiaceae: 1 (1968); R. A. Dyer, Gen. 2: 912 (1976). Type species: *M. hastata* Solms-Laub.

Perennial aquatic herbs, erect, usually deciduous. *Leaves* radical, emergent, petiolate with cordate to ovate blade. *Flowers* racemose or subumbellate, pedicelled. *Perianth-lobes* 6, free, subequal, blue, marcescent. *Stamens* 6, arising from base of segments, unequal, one larger with an erect apical tooth on one side of filament; anthers erect, basifixed, opening by apical slits. *Ovary* 3-locular with axile placentas, many-ovulate; style filiform, stigma apical, 3-lobed, small. *Capsule* spindle-shaped, with thin pericarp enclosed by marcescent perianth; seeds numerous, minute, ovoid, with several thin ribs.

A genus of 7 species, occurring in Africa, Asia and Australia. One tropical African species recorded from the Transvaal Lowveld.

The name *Monochoria* alludes to the one stamen being different; it is larger, blue, and its filament bears an apical upright tooth.

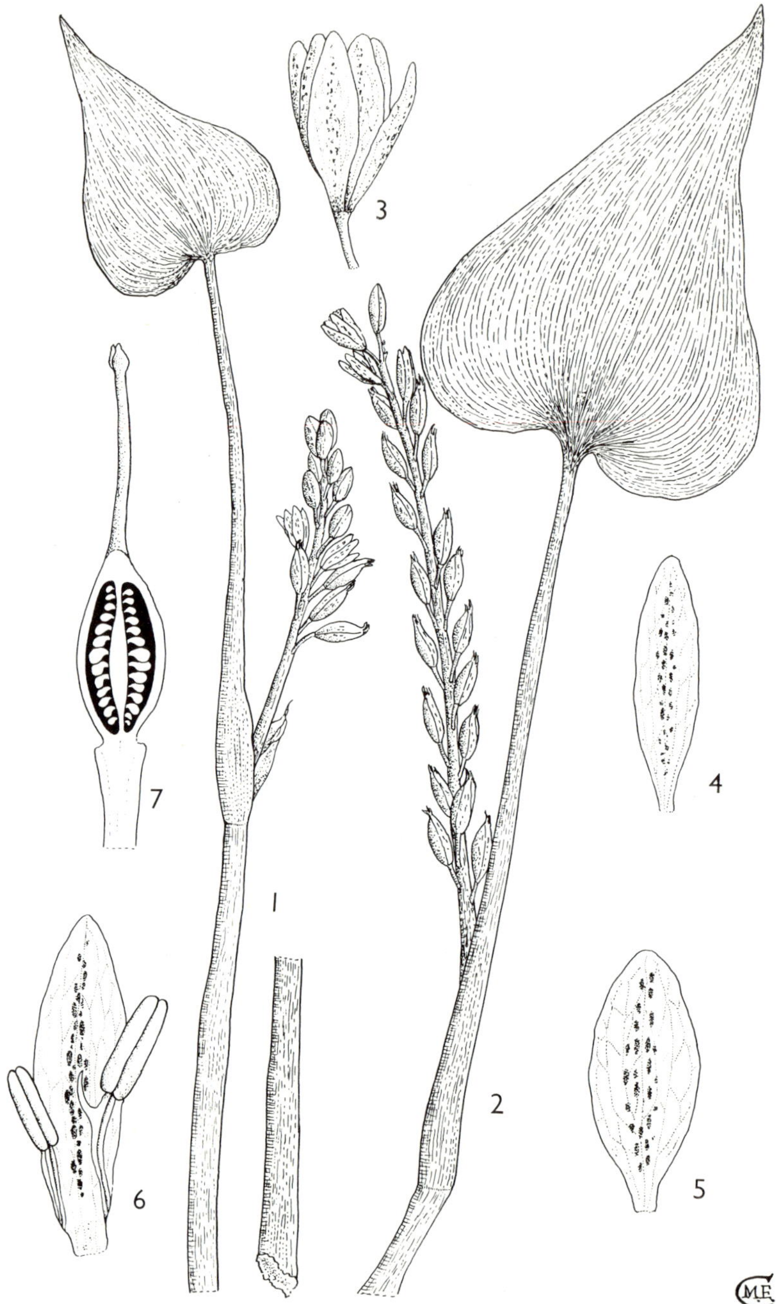

Fig. 18.—**Monochoria africana**: 1 & 2, flowering stems, × 0,6; 3, flower, × 3; 4, outer perianth-lobe, × 3; 5, inner perianth-lobe, × 3; 6, perianth-lobe with one each of the two types of stamens attached, × 4; 7, gynoecium in longitudinal section, × 6 (all from *Greenway & Rawlins* 9483). Reproduced from the 'Flora of Tropical East Africa' with permission of the Director of the Royal Botanic Gardens, Kew.

Monochoria africana *(Solms-Laub.) N.E. Br.* in F.T.A. 8: 5 (1901); Schwartz in Bot. Jb. 61, Beibl. 139: 36 (1927); F. W. Andr., Fl. Pl. Anglo-Egypt. Sudan 3: 278 (1956); Verdc. in Kirkia 1: 81, t. 8 (1960), F.T.E.A. Pontederiaceae: 3, fig. 1 (1968); Berhaut, Fl. Seneg. 316, t. opp. p. 448 (1967). Type: Sudan, Equatoria, Jur Ghattas, *Schweinfurth* 2296 (B, holo.†; K; PRE!).

M. vaginalis (Burm. f.) Presl ex Kunth var. *africana* Solms-Laub. in A. DC., Monogr. Phan. 4: 525 (1883).

Tufted aquatic annual up to 0,75 m tall, leaves and racemes basal, roots fibrous, very numerous. *Leaves:* lamina cordate to ovate, c. 100 mm long, attenuated into acute apex; petioles up to 70 mm long, sheathed basally. *Flowering stem* bearing simple, shortly pedunculate raceme, emerging from leaf-sheath of somewhat reduced apical leaf; rhachis up to 200 mm long, many-flowered, pedicels thin, c. 10 mm long. *Flowers* with segments nearly free to base, c. 15 mm long, blue with red glands dotted along midrib. *Stamens* somewhat unequal; 5 smaller, yellow, 6th larger with blue anther. *Ovary* oblong-globose; style somewhat declinate; stigma papillate. *Capsule* fusiform, 14 mm long; permanent style form- ing beak, enclosed in marcescent perianth, dehiscing explosively; seed typical.

The only known record for the region is from Transvaal in the southern part of the Kruger National Park, where it occurs along the sides of shallow seasonal pools from November to May. Widespread in tropical and subtropical Africa but rarely collected. Plants (excluding roots) edible. Map 51.

Vouchers: *Stevenson-Hamilton* in PRE 2281; *Van Wyk* 4662.

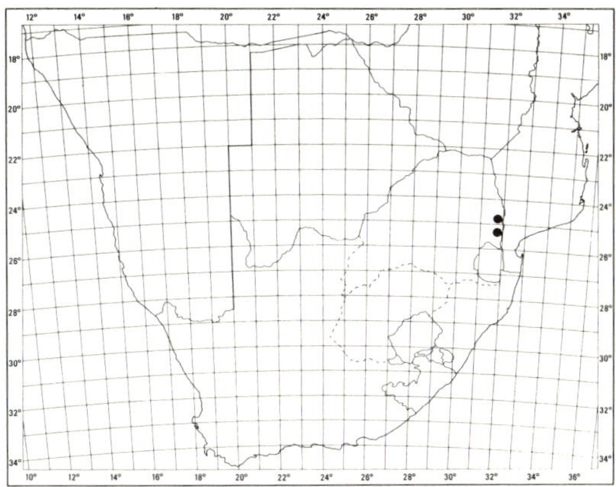

MAP 51.— **Monochoria africana**

921   2. EICHHORNIA

Eichhornia *Kunth*, Eichhornia gen. nov. fam. Pontederiaceae (1842), and in Enum. Pl. 4: 129 (1843); Solms in A. DC., Monogr. Phan. 4: 525 (1883); N. E. Br. in F.T.A. 8: 4 (1902); Podlech in F.S.W.A. 154: 1 (1969); R. A. Dyer, Gen. 2: 912 (1976), nom. cons. Type species: *E. azurea* (Swartz) Kunth.

Perennial aquatic herbs, rooting in mud or free-floating. *Roots* many, long and feathery in floating plants. *Leaves* dimorphous; emergent ones often with swollen, long petioles and obovate, orbicular, spathulate or lanceolate blades; submerged ones linear, with petioles sheathing stems, shortly stipulate. *Flowers* in well developed panicles or spikes or these reduced to 2–1 flowers. *Perianth* funnel-shaped, with a somewhat curved tube and 6 subequal spreading lobes, blue or mauve. *Stamens* 6, upper included, lower exserted, anthers dorsifixed, oblong, opening lengthwise. *Ovary* 3-locular; ovules numerous; style filiform; stigma slightly dilated, entire or very shortly lobed. *Capsule* spindle-shaped, covered with marcescent perianth. *Seeds* numerous, small, ovoid, finely ribbed. *Chromosomes:* n = 16.

A genus of 5 American species, but one of these *(E. natans,)* also occurs wild in northern South West Africa/Namibia and further north in tropical Africa. *Eichhornia crassipes* from Brazil is now naturalized in Southern Africa and elsewhere, and is a widespread noxious weed.

Named after an eminent Prussian Minister, J. A. Eichhorn, 1779–1856.

1 Leaves dimorphous: submerged leaves linear and cordate, emergent leaves arranged on long spreading stems; perianth-limb c. 10 mm in diam. .... 1. *E. natans*
1 Leaves uniform, clustered, emergent, with a spade-shaped lamina and swollen petiole; perianth-limb c. 50 mm in diam. .... 2. *E. crassipes*

1. **Eichhornia natans** *(P.Beauv.) Solms-Laub.* in Abh. naturw. Ver. Bremen and in A. DC., Monogr. Phan. 4: 526 (1883); N. E. Br. in F.T.A. 8: 4 (1901); Hepper in F.W.T.A. edn 2, 3,1: 110, fig. 356 (1968); Podlech in F.S.W.A. 154: 2 (1969). Type: Nigeria, Oware, banks of Formosa River, *Palisot de Beauvois* s.n. (G, holo.).

Pontederia natans P.Beauv., Fl. Owar. 2: 18, t. 68, fig. 2 (1810).

Herbaceous rooted aquatic forming dense, submerged mats, upper branches with emergent leaves and flowers. *Stems* long, thin, rooting at lower nodes, roots long, with numerous thin rootlets. *Submersed leaves* thin, linear, up to 80 × 1–3 mm, sheathing at base, sheath auriculate. *Floating leaves* with long petiole; lamina cordate, c. 10–20 × 10–15 mm, entire, upper surface minutely pustulate. *Flowers* terminal, solitary, enclosed below by a tubular spathe with its short apex abruptly recurved, apiculate; perianth-tube narrowly cylindrical, up to 20 mm long; lobes spreading, narrowly ovate, c. 6 mm long; mauve or white. *Stamens* free. *Capsule* fusiform, attenuated into a long, persistent style, c. 15 mm long. *Seeds* numerous, typical. Fig. 19.

Recorded from South West Africa/Namibia and Botswana, widespread in subtropical and tropical Africa; also in tropical America and Cuba. Map 52.

Vouchers: *De Winter & Marais* 5032; *Dinter* 7247; *Gibbs Russell & Biegel* 1536; *Smith* 399.

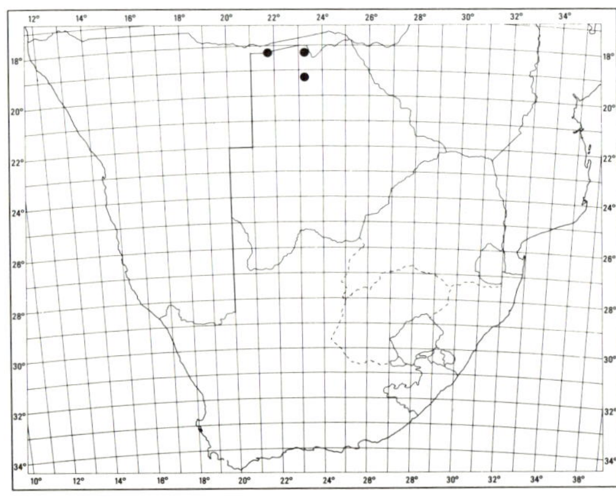

MAP 52.— **Eichhornia natans**

2. **Eichhornia crassipes** *(Mart.) Solms-Laub.* in A. DC., Monogr. Phan. 4: 527 (1883); Schwartz in Natürl. PflFam. edn 2, 15a: 186 (1930); Adamson in Adamson & Salter, Fl. Cape Penins. 160 (1950); Backer in Van Steenis, Fl. Males. ser. 1,4: 259, figs 2, 3 (1951); Henderson & Anderson in Mem. bot. Surv. S. Afr. 37: 66, 67 (1966); Sculthorpe, Biol. Aquat. Vasc. Plants: 280, 460, etc. (1967); Verdc. & Carter, in F.T.E.A. Pontederiaceae: 4 (1968). Type: Brazil, Minas Geraes, *Martius* s.n. (M, holo.).

Pontederia crassipes Mart., Nov. Gen. Sp. Pl. 1: 9, t. 4 (1824). *E. speciosa* Kunth, Enum. Pl. 4: 131 (1843).

Pontederia azurea sensu Hook. in Curtis's bot. Mag. t. 2932 (1829), non Sweet.

Herb, free-floating or rooted in mud, 40–400(–800) mm tall. *Roots* many, long, densely feathery. *Leaves* 4–8, clustered on short rhizomes, erect, petiole and lamina very variable in size and shape: in young, free-floating plants petiole 50–100 mm long, swollen and fusiform in lower half, narrowly cylindrical above; blade spade-shaped or obtusely rounded, 40–90 mm wide, smooth, firm; in older large plants rooted in mud or under crowded conditions, petiole cylindrical, up to 500 mm long and lamina ovate, up to 100 mm long. *Inflorescence* a c. 8-flowered showy spike, raised above leaves. *Perianth* with short, curved tube, c. 20 mm long; limb subbilabiate, c. 50 mm in diam., delicate, fugaceous, pale mauve, central upper lobe broadest with large blue area and yellow spot in centre. *Stamens* dimorphous, lower 3 exserted, with glandular-hairy filaments curved upwards at apex; upper 3 short, reaching mouth of tube. *Ovary* ovoid, ovules numerous; style in Southern African plants c. 25 mm long, shorter than long stamens, glandular-hairy; stigma globose, glandular-lamellate.

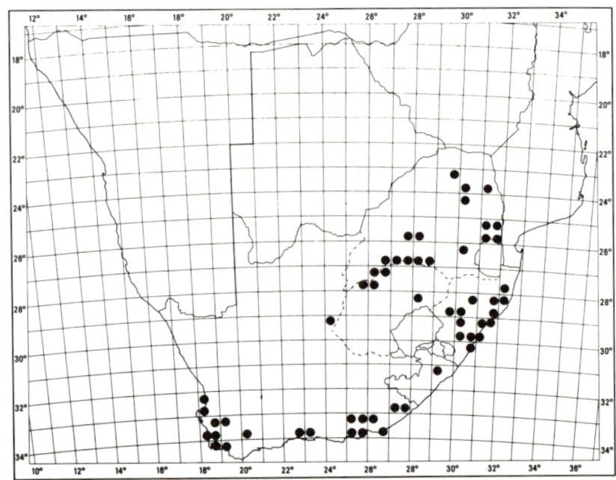

MAP 53.— **Eichhornia crassipes**

Fig. 19.—**Eichhornia natans**: 1, flowering stem, × 1; 2, flower, × 4; 3, gynoecium, × 4; 4, tip of style and stigma, × 16; 5, capsule and part of spathe, × 2; 6, part of capsule, with wall partly removed to show seeds, × 8; 7, transverse section of capsule, oblique view, × 8; 8, seed, × 32 (all from *Duke* 2). Reproduced from the 'Flora of Tropical East Africa' with permission of the Director of the Royal Botanic Gardens, Kew.

Capsule 3-locular, splitting longitudinally, becoming mucilaginous and disintegrating in water when ripe; seed apomictic, rarely produced, terete, truncate, c. 12-ribbed, rarely germinating.

Vouchers: *Crawford* 560; *Musil* 351; *Ward* 8926.

The water hyacinth, a native of South America, now widespread in all tropical and subtropical countries, was introduced here in 1884 and has now become naturalized all over the region. It is a menace in most quiet waters and control has proved very costly. Flowers from this country all possess 3 stamens which are longer than the style but it was seen that in some Zimbabwean collections the style was exserted beyond the stamens. (Lomagundi district, Hunyani River, *Phipps* 2491, *Jacobsen* 2433; see also Wild, Harmful Aquatic Plants, Africa and Madagascar, t. 1; 1961). Schürhoff in Ber. dt. bot. Ges. 40: 60 (1922) established that the pollen is sterile, the seeds developing apomictically (agamospermy). Map 53.

922   **3. PONTEDERIA**

Pontederia *L.*, Sp. Pl. 288 (1753); Benth. & Hook. f., Gen. Pl. 3, 2: 837 (1883); Schwartz in Natürl. PflFam. edn 15a: 188 (1930); Rendle, Class. Flow. Pl. 1: 281 (1953); Lowden in Rhodora 75: 426–487 (1973). Type species: *P. cordata* L.

Perennial herbaceous hygrophytes, terrestrial or free-floating. *Leaves* usually exserted. *Inflorescence* with numerous sessile, clustered flowers forming a dense blue or rarely white spike. *Flowers* with perianth forming a tube and limb somewhat bilabiate, middle lobe of upper 3 wider and with a central blotch; lower 3 similar; lower part of perianth persistent, enveloping utricle. *Stamens* 6, unequal, anther-locules opening by slits. *Ovary* 1-locular; ovule 1, pendulous; style slender, short, medium or long; stigma apical. *Utricle* winged. *Chromosomes:* n = 8.

An American genus with 5 species. The species below has become naturalized in some areas of the Republic.

Named after Guilio Pontedera, a professor of Botany in Padua, 1688–1757.

Pontederia cordata *L.*, Sp. Pl. 288 1753); Lam., Tabl. Encycl. 2, t. 225 (1797); Ker-Gawl. in Curtis's bot. Mag. 29, t. 1156 (1808); Solms-Laub. in A. DC., Monogr. Phan. 4: 532 (1883); R. W. Sm. in Bot. Gaz. 25: 324 (1898), l.c. 45: 338 (1908); Coker, l.c. 44: 293 (1907); Schwartz in Natürl. PflFam. edn 2, 15a: 188 (1930). Type: LINN 407.42 (LINN, holo.; photo.!).

var. **ovalis** *(Mart.) Solms* in A. DC., Monogr. Phan. 4: 533 (1883). Type: Brazil, *Martius* s.n. (M, holo.).

P. ovalis Mart. in Roem. & Schult., Syst. Veg. 7: 1140 (1830).

Rooted herbaceous hygrophyte 1–2 m tall, with a horizontal rhizome, forming colonies. *Stems* long, sympodial, 1-leaved, leaves erect, exserted, with short petiole forming dilated sheath at base; lamina cordate (in typical variety) to ovate or 'lanceolate' (in var. *ovalis*), c. 230 × 70 mm, firm, smooth. *Flowering stem* emerging from sheathing base of leaf and about as long; peduncle short, pubescent, enveloped by large sheathing bract. *Spike* cylindrical, up to c. 100 mm long and 20 mm broad, flowers dense, clustered, blue, rarely white. *Perianth* c. 15 mm long, shortly pubescent on outside, funnel-shaped, upper lip broader than side-lobes, with yellow blotch in centre. *Stamens* unequal, upper shorter, included, lower exserted, filaments filiform, pubescent. *Ovary:* abortive carpels surviving as ridges on walls of ovary; style long or short, slender; stigma apical. *Utricle* enveloped by accrescent base of perianth. Fig. 20.

South America, naturalized in North America. A well known decorative, cultivated aquatic, common in parks and gardens, recently found growing wild along riverbanks in Kwazulu; also said to occur in the eastern Cape. Common name in America: Pickerell Weed. Map 54.

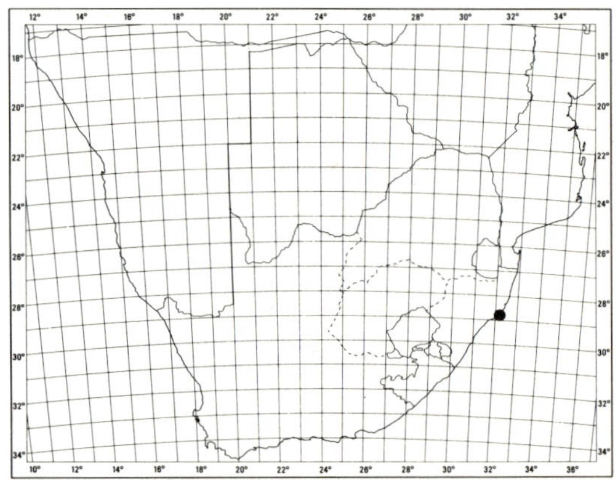

MAP 54.— **Pontederia cordata**

Fig. 20.—**Pontederia cordata**: a, flowering stem, × 0,5; b, flower, × 2,5 (*Mauve* 5442).

Voucher: *Musil* 124.

R. W. Smith in Bot. Gaz. 25: 324 (1898) observed that the pollen of this species is sterile. Coker in Bot. Gaz. 44: 293 (1907) described the development of the seed.

Both var. *cordata* and var. *ovalis* have been introduced into Southern Africa. So far only var. *ovalis* has been found growing wild in Kwazulu. The typical variety, like var. *ovalis*, has pubescent peduncles, a fact not mentioned by Lowden (1973).

924

4. HETERANTHERA

Heteranthera *Ruiz & Pav.*, Fl. Peruv. chil. Prodr. 9, t. 2 (1794); N. E. Br. in F.C. 7: 1 (1897); Verdc. in F.T.E.A. Pontederiaceae: 6, t. 3 (1968); Podlech in F.S.W.A. 154: 2 (1969); R. A. Dyer, Gen. 2: 913 (1976); nom. cons. Type species: *H. reniformis* Ruiz & Pav.

Perennial aquatic herbs, rhizomatous, rooting in mud. *Leaves* with long, swollen, spongy petiole with expanded sheathing base; blade cordate-hastate or ovate, floating or occasionally all leaves submerged, linear. *Flowers* small, in a terminal spike, shortly raised above surface of water, rarely reduced to 1–2, occasionally some cleistogamous. *Perianth* regular or nearly so, salver- or funnel-shaped, blue, mauve, violet or white, accrescent. *Stamens* 3 (1 in cleistogamous flowers), exserted; filaments arising from throat of tube; median stamen occasionally longer; anthers dorsifixed, opening by longitudinal slits. *Ovary* imperfectly 3-locular with 3 parietal, multi-ovulate placentas; style filiform; stigma small. *Capsule* oblong or terete; seeds numerous, ovoid.

A genus of about 10 species, mostly American; also in tropical Africa; one of these found as far south as South West Africa/Namibia and Transvaal. The name refers to the dimorphous anthers.

Heteranthera callifolia *Kunth*, Enum. Pl. 4: 121 (1843); N. E. Br. in F.C. 7: 1 (1897), and in F.T.A. 8: 2 (1901); Hepper in F.W.T.A. edn 2, 3: 111 (1968); Verdc. in F.T.E.A. Pontederiaceae: 6, fig. 3 (1968); Podlech in F.S.W.A. 154: 2 (1969). Type: Senegal, *Sieber* 51 (P, holo.; K).

H. kotschyana Fenzl ex Solms-Laub. in Schweinf., Beitr. Fl. Aethiop. 1: 205 (1867); N. E. Br. in F.C. 7: 2 (1897). Type: Sudan Republic, Kordofan, by Mulbes at Obeid, *Cienkowsky* 378 (W, holo.).

Somewhat fleshy, creeping aquatic herbs up to c. 0,3 m tall, rooting at nodes, roots densely covered with short rootlets; exposed parts dying back in dry season, rhizomes probably perennial. *Leaves* with erect, hollow petioles up to c. 200 mm long, and cordate, floating blades c. 50 mm long; old leaf-sheaths purple-tinged. *Spikes* up to 10-flowered, minutely glandular-pubescent, just raised above water. *Normal flowers* subregular; perianth-tube c. 10 mm long, lobes 6, more or less equal, narrowly ovate, 5 mm long, white, blue, mauve or violet. *Stamens* exserted from tube placed on anterior side; filaments inserted on tube, white or blue; anthers yellow, reflexed, dorsifixed near base. *Ovary* ovoid; style deflected to one side, short; stigma small. *Capsule* cylindrical, apiculate, c. 10 mm long, ensheathed by accrescent perianth-tube; walls disintegrating at ma-

turity; seeds typical. *Cleistogamous flower* solitary, hidden inside spathe of spike, which in turn is subtended by leaf-sheath. *Capsule* c. 15 mm long, larger and with more seeds than in that of normal flowers. *Buds* in lower leaf-axils occasionally forming vegetative shoots. Fig. 21.

Recorded from South West Africa/Namibia, Botswana and the Transvaal. Widespread in subtropical and tropical Africa. In swampy areas, vleis, pans or rock pools which become dry in winter. Map 55.

Vouchers: *Dinter* 7353; *Germishuizen* 36; *Schoenfelder* S814; *Smith* 496; *Theron* 2980.

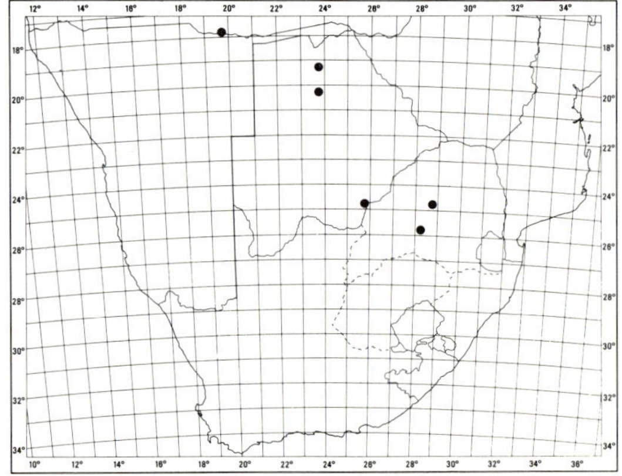

MAP 55.— **Heteranthera callifolia**

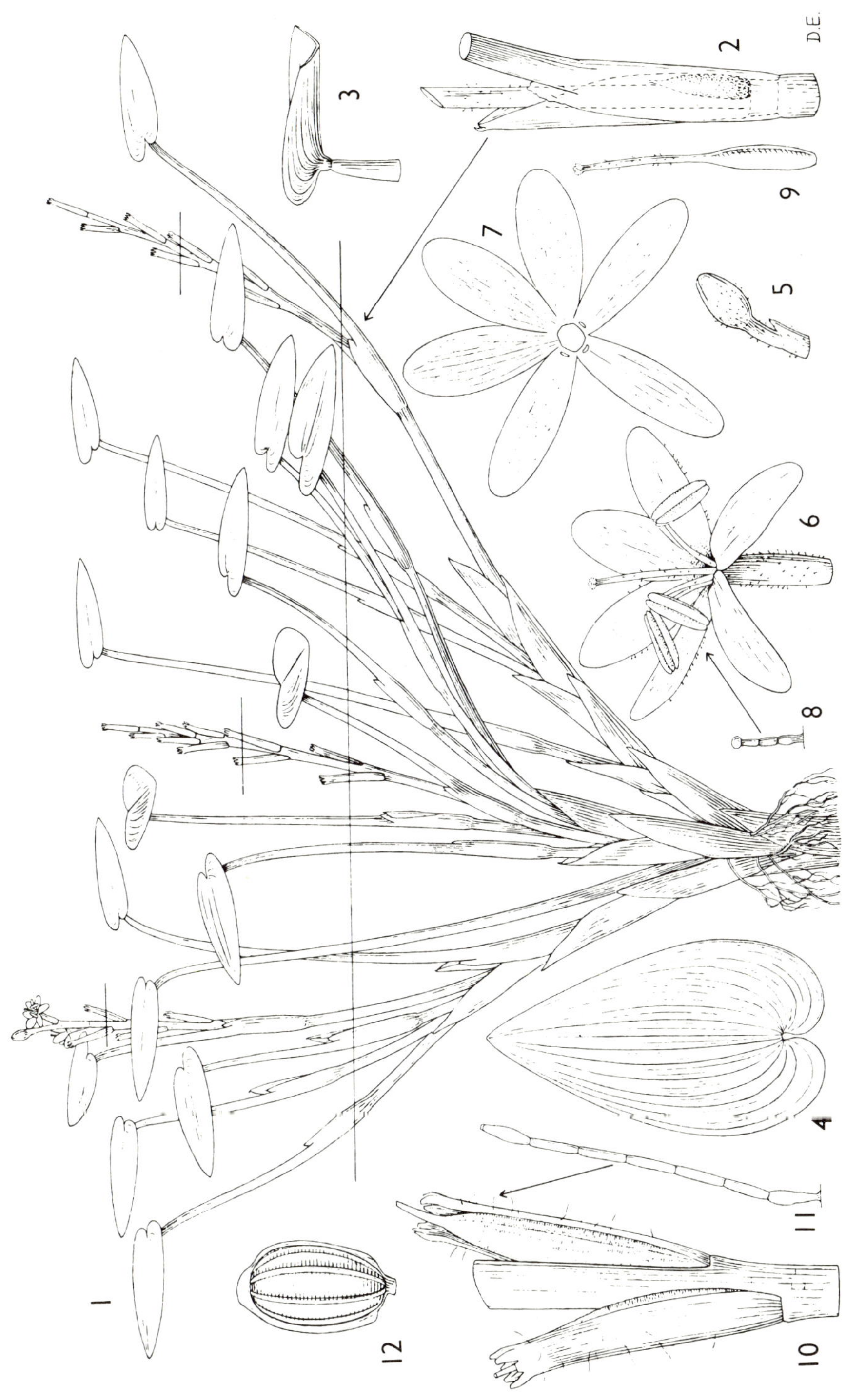

FIG. 21.—**Heteranthera callifolia**: 1, habit, × 0,5; 2, detail of leaf-sheaths, × 1,5; 3, upper part of petiole and base of leaf-blade, × 1; 4, leaf-blade, surface view, × 1; 5, bud, × 4; 6, flower, × 4; 7, perianth, detached and flattened, × 4; 8, glandular hair from perianth, × 40; 9, gynoecium, × 4; 10, part of infructescence with two capsules, × 2; 11, glandular hair from perianth remnants sheathing capsule, × 40; 12, seed, × 22 (1–4, 10–12, from *Polhill & Paulo 1704*; 5–9, from *Polhill & Paulo 2168*). Reproduced from the 'Flora of Tropical East Africa' with permission of the Director of the Royal Botanic Gardens, Kew.

JUNCACEAE

by A. A. OBERMEYER

Annual or perennial rhizomatous hygrophytic herbs often with aerenchymatous tissue, or with a woody stem bearing an apical leaf-rosette *(Prionium)*. *Leaves* grass-like, or tubular and pungent, sometimes septate, lower ones often reduced to cataphylls; or in *Prionium* ovate-acuminate, hard, spinous. *Flowers* bracteate, on long, usually naked peduncles, in terminal or pseudolateral panicles (anthelae) or congested into capitula. *Perianth* small, regular, bisexual, marcescent. *Tepals* with 3 outer and 3 inner subsimilar, glumaceous, green or brown. *Stamens* (3)6, hypogynous; filaments filiform or trigonous; anthers 2-thecous, basifixed, introrse, dehiscing longitudinally; pollen in tetrads. *Ovary* 1- or 3-locular; ovules 3–many, biseriate, basal, axile or parietal; style present or 0; stigmas 3. *Capsule* 1–3-locular, dehiscing loculicidally; seeds very small, 3–many, ovoid, obovoid, globose, spindle-shaped or compressed, often with basal and/or apical appendages, occasionally becoming mucilaginous.

Genera 8, species about 300, cosmopolitan, in temperate or cold regions; 3 genera in Southern Africa; usually hygrophytes, anemophilous or dispersed by animals (in cases where the seeds become mucilaginous).

1 Robust perennial, caulescent plants with an apical leaf-rosette; leaves serrate   1. **Prionium**
1 Rushes, perennial or annual, with basal, entire leaves:
   2 Leaves glabrous; leaf-sheaths usually open; flowers in decompound inflorescences to much reduced single capitula; capsules many-seeded .... 2. **Juncus**
   2 Leaves with long, soft, inconspicuous, scattered hairs; leaf-sheaths closed; flowers in an oblong, compact head; capsules 3-seeded .... 3. **Luzula**

930   **1. PRIONIUM**

Prionium *E. Mey.* in Linnaea 7: 130 (1832); Buchen. in Abh. naturw. Ver. Bremen 4: 408 (1875) and in Pflanzenreich 4, 36 (Heft 25): 41 (1906); Bak. in F.C. 7: 28 (1897); Adamson in Adamson & Salter, Fl. Cape Penins. 161 (1950); Cutler in Anat. Monocot. 4: 65–69, etc. (1969); R. A. Dyer, Gen. 2: 914 (1976). Type species: *Prionium serratum* (L. f.) Drège ex E. Mey.

Robust stoloniferous perennials with leaves in dense apical rosettes. *Stems* stout, erect, branched only basally, densely covered with black fibrous remains of old leaves. *Leaves* long, rigid, with tubular sheathing base. *Inflorescence* a large, much branched terminal panicle; peduncle exserted, trigonous; side branches and branchlets bearing few- to many-flowered fascicles; bracts funnel-shaped, caudate. *Flowers* shortly pedicelled. *Tepals* sub-equal, ovate, rigid, brown. *Stamens* 6, usually exserted; filaments filiform; anthers basifixed, oblong, with opposing locules, opening lengthwise. *Ovary* ovoid, 3-locular; ovules 3–6 in lower half of each locule, axile; style short or 0; stigmatic branches 3, thick, papillate. *Capsule* obovoid, tricostate; seeds 1(–2) per locule, ovoid-oblong.

Species 1, endemic in the Cape and southern Natal.

Cutler in Anat. Monocot. 4: 65–69, etc. (1969) would prefer to place this genus in a family on its own.

The name *Prionium* is derived from the Greek *prion*, a saw.

Prionium serratum *(L. f.) Drège ex E. Mey.* in Drège, Zwei Pfl. Doc. 10 (1843); Hook. in Hooker, Lond. J. Bot. 9: 173 (1857); Hook. f. in Curtis's bot. Mag. t. 5722 (1868). Type: Cape, *Thunberg*, LINN 449: 50 (LINN, holo., PRE, photo.!).

Juncus serratum L. f., Suppl. 208 (1782).

Prionium palmita E. Mey. in Linnaea 7: 131 (1832); Kunth, Enum. 3: 315 (1841); Bak. in F.C. 7: 28 (1897). Type: Cape, Table Mountain, *Ecklon* (S, holo.).

Stems up to about 2 m high, 50–100 mm in diameter. *Leaves* linear-acuminate, up to 1

FIG. 22.—**Prionium serratum**: 1, habit (much reduced, after photo, *Strey* 8288); 2, flowering twig, × 0,4 (*Strey* 8288); 3, fibrous remains of leaf base spread out, × 0,6; 4, part of leaf showing serrate margin, × 1,2; 5, transverse section of leaf showing fibrovascular bundles, × 3,5; 6, ripe fruit, carpels dehiscing at ventral suture, showing seeds, × 4; 7, flower, × 3; 8, ovary, × 3; 9, ovary (transverse section), × 7; 10, seed with loose envelope, × 7,6; 11, seed after removal of envelope, × 7,6 (figures 3–6; 9 & 10 after Marloth, The Flora of South Africa 4: 72, fig. 19; 7 & 8 after *Strey* 7764; 11 after *Marloth* 467).

m long and up to 80 mm broad near the base, margins sharply serrulate. *Inflorescence* erect, up to 0,5 m high; branches ascending, c. 0,2 m long, bearing numerous flowering branchlets c. 40 mm long. *Flowers* in lateral fascicles surrounded below by funnelform, caudate bracts, diminishing in size towards the apex. *Perianth* segments acute, c. 3 mm long, bright brown. *Stamens* as long as perianth or slightly longer. *Ovary* ovoid, obtusely 3-angled. Fig. 22.

An aquatic or semi-aquatic, common in the southern Cape, often choking rivers; along the coast and further inland as far as Tulbagh and Ceres; in the S.E. Cape it has been collected at Howieson's Poort, and in the Transkei at Lusikisiki. Frequent also in southern Natal, along the coast and further inland. Map 56.

Vouchers: *Codd* 9704; *Galpin* 4780; *Marloth* 467, 5164b; *Mauve & Hugo* 121; *Miller* 154; *Schonland* 3306; *Strey* 7764; 8288.

The leaves contain strong fibres and were formerly used for plaiting straw hats. Common name: Palmiet.

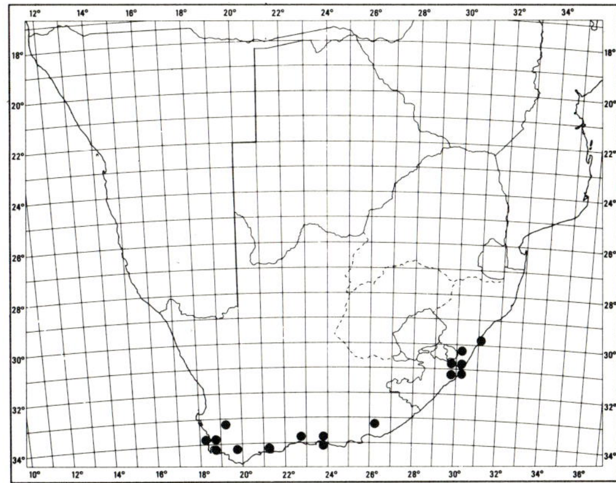

MAP 56.— **Prionium serratum**

936   **2. JUNCUS**

Juncus *L.*, Sp. Pl. 325 (1753); Buchen. in Abh. naturw. Ver. Bremen 4: 393 (1875), and in Bot. Jb. 12: 167 (1890), and in Pflanzenreich 4, 36 (Heft 25): 98 (1906); Bak. in F.C. 7: 17 (1897), and in F.T.A. 7: 92 (1897); Vierhapper in Natürl. PflFam. edn 2, 15a: 214 (1930); Adamson in J. Linn. Soc., Bot. 50: 1 (1935); Weim. in Svensk bot. Tidskr. 40: 141 (1946); Adamson in Adamson & Salter, Fl. Cape Penins. 161 (1950); Snogerup, Flora Iran. Juncaceae: 1 (1971); R. A. Dyer, Gen. 2: 914 (1976). Type species: *Juncus acutus* L.

Annuals or perennials, stiff and pungent, rush- or grass-like, rhizomatous, usually tufted, glabrous. *Roots* thin and numerous or, in perennials growing in saline habitats, thick and woody, covered by a dense velamen of root-hairs. *Leaves* basal or on stolons, many to few–1 per shoot, flat and soft or cylindric, rigid and often pungent; in subgenus *Septati* transversely septate; lower leaves often reduced to cataphylls; sheaths open, rarely cylindrical at first, often long and auriculate. *Flowers* proterogynous, occasionally cleistogamous, or turning into vegetative buds which form new plants; actinomorphic, subtended by 3 or 1 prophylls; arranged in much branched dichasia or reduced to a solitary capitulum; the flowers in clusters, capitula or solitary but congested on secund, spicate branchlets. *Tepals* glumaceous, narrowly ovate, usually with a broad firm midrib and membranous sides, outer usually longer, enveloping inner, often aristate; inner shorter, often broader, rarely enlarging in fruit, apex acute or aristate. *Stamens* 6, or rarely 3 inner aborted, hypogynous, shorter than tepals (in S. African species); filaments filiform, very short or longer; anthers basifixed, introrse, bilocular, dehiscing longitudinally; pollen in tetrads. *Ovary* unilocular initially (and remaining so in subgenus *Septati*), becoming 3-locular in most species by centripetal development of placental ridges meeting in centre to form an axis; style long to short or 0; stigmas 3, short and often convolute, to very long and exserted, often red, with long glutinous papillae. *Capsule* 1–3-locular or triseptate, loculicidal with placental axis splitting down centre or rarely remaining intact; seeds minute, very numerous, globose to ellipsoid or obconic, one side sometimes flattened, reticulate or striate, scobiform or smooth, in some species covered with a thin hygroscopic membrane.

Species about 250; cosmopolitan but predominantly in temperate climates. In Southern Africa 21 species; in wet surroundings such as riverbanks, marshes, sandy beaches, saltpans, or in temporary moist depressions along roads, fields and open spaces, often covering vast areas. Flowering usually in spring.

The name *Juncus* is derived from the Latin *'jungere'* meaning to join, and refers to its use in ancient times for producing plaited mats, chairseats, etc. The pith was used as a wick. It is also suitable for paper making.

1 Each flower subtended by a bract and 2 bracteoles:
  2 Leaves flat or semiterete, grooved; flowers in divergent or unilateral, secund spikes; cataphylls 0 or small (subgenus *Poiophylli* Buchen.):
    3 Annuals, small; stems leafy, all flowering freely; lamina flat, grass-like; without cataphylls ...... 1. *J. bufonius*
    3 Perennials; leaves at or near base, flat or semiterete; flowers on bare scapes; cataphylls present:
      4 Leaves grass-like, flat; flowers in loose terminal panicles; tepals spreading, overtopping capsules   2. *J. tenuis*
      4 Leaves semiterete, sulcate, firm; flowers pseudolateral, subtending bract forming continuation of stem; tepals about as long as capsules:
        5 Plants up to 0,6 m tall, sclerotic; tepals 5 mm long; flowering in spring .... 3. *J. imbricatus*
        5 Plants up to 0,25 m tall, softer; tepals 3 mm long; flowering in autumn .... 4. *J. capillaceus*
  2 Leaves cylindric, cauliform, pungent, surrounded below by sheathing cataphylls; inflorescence pseudolateral, subtending bract forming continuation of stem (subgenus *Genuini* Buchen.):
    6 Stamens 3; tepals 2 mm long, spreading; pith continuous; capsule obtuse, thin .... 5. *J. effusus*
    6 Stamens 6; tepals 4 mm long, erect; pith with lacunae; capsule acute, hard .... 6. *J. inflexus*
1 Each flower subtended by a single bract:
  7 Leaves few, cylindric, hard and pungent or softer and acute; cataphylls well developed, sheathing; perennials:
    8 Leaves non-septate, basal, rigid, pungent; flowers clustered in compound, pseudolateral inflorescences; seeds appendaged or apiculate (subgenus *Thalassici* Buchen.):
      9 Capsule globose, exserted from perianth; inner tepals with apical hyaline auricles .... 7. *J. acutus*
      9 Capsule subterete, triangular, about as long or just exserted from perianth; inner tepals with a straight, narrow, hyaline margin:
        10 Inflorescence usually plumose, compact, flowers in glomerules; capsule as long as perianth; seed apiculate; a coastal species, rarely inland .... 8. *J. kraussii* subsp. *kraussii*
        10 Inflorescence more loosely branched, with flowers 1–few, on shorter and longer side branchlets; capsule exserted; seed appendaged; in brackish areas or saltpans in the interior .... 9. *J. rigidus*
    8 Leaves septate, cauline, fairly soft and pointed; flowers in capitula on divaricate terminal inflorescences; seeds apiculate (subgenus *Septati* Buchen.):
      11 Flowering stem with 1 long leaf; capsule tri-locular; medulla of stem consisting of a central and lateral airspaces .... 10. *J. punctorius*
      11 Flowering stem with 3–5 leaves; capsule unilocular; medulla of stem with a central cavity only:
        12 Capsule ovoid, obtuse, enclosed in perianth .... 11. *J. oxycarpus*
        12 Capsule cylindrical, beaked, exserted from perianth .... 12. *J. exsertus*
  7 Leaves many, flat or channelled, basal, not septate; cataphylls absent; flowers in capitula; perennials or annuals with many thin roots (subgenus *Graminifolii* Buchen.):
    13 Perennials; flowers in terminal panicles; capitula dense to few-flowered:
      14 Leaves broad, soft, flat, rosulate; plants repent .... 13. *J. lomatophyllus*
      14 Leaves grass-like, linear-acuminate to filiform; plants erect, tufted:
        15 Style and stigmas long, red, much exserted; anthers longer than short filaments; perianth 4–5 mm long .... 14. *J. capensis*
        15 Style 0 or very short; stigmas spreading on ovary; anthers short, as long as filaments; perianth 3 mm long .... 15. *J. dregeanus*
    13 Annuals; capitula solitary or in panicles, many- to few-flowered:
      16 Stamens 6:
        17 Capitula solitary or rarely with 1–2 smaller younger ones, brush-like; plants scabrid; capsule cylindrical .... 19. *J. scabriusculus*
        17 Capitula in panicles:
          18 Anthers small, as long as filaments; style and stigmas short; capitula small, few-flowered, cup-shaped; thin wiry plants .... 18. *J. rupestris*
          18 Anthers long on very short filaments; style and stigmas long, exserted; capitula many-flowered:
            19 Capsule shortly beaked; capitula with flowers closely pressed together; tepals aristate; inner acute, enlarging with age, outstripping outer .... 16. *J. cephalotes*
            19 Capsule with a long beak; capitula with narrow flowers squarrosely arranged, shortly pedicelled; tepals long, aristate .... 17. *J. stenopetalus*
      16 Stamens 3, rarely 6; plants with capitulum pseudolateral:
        20 Subtending bract exserted, forming continuation of stem; cosmopolitan .... 20. *J. capitatus*
        20 Subtending bract very short; Namaqualand .... 21. *J. obliquus*

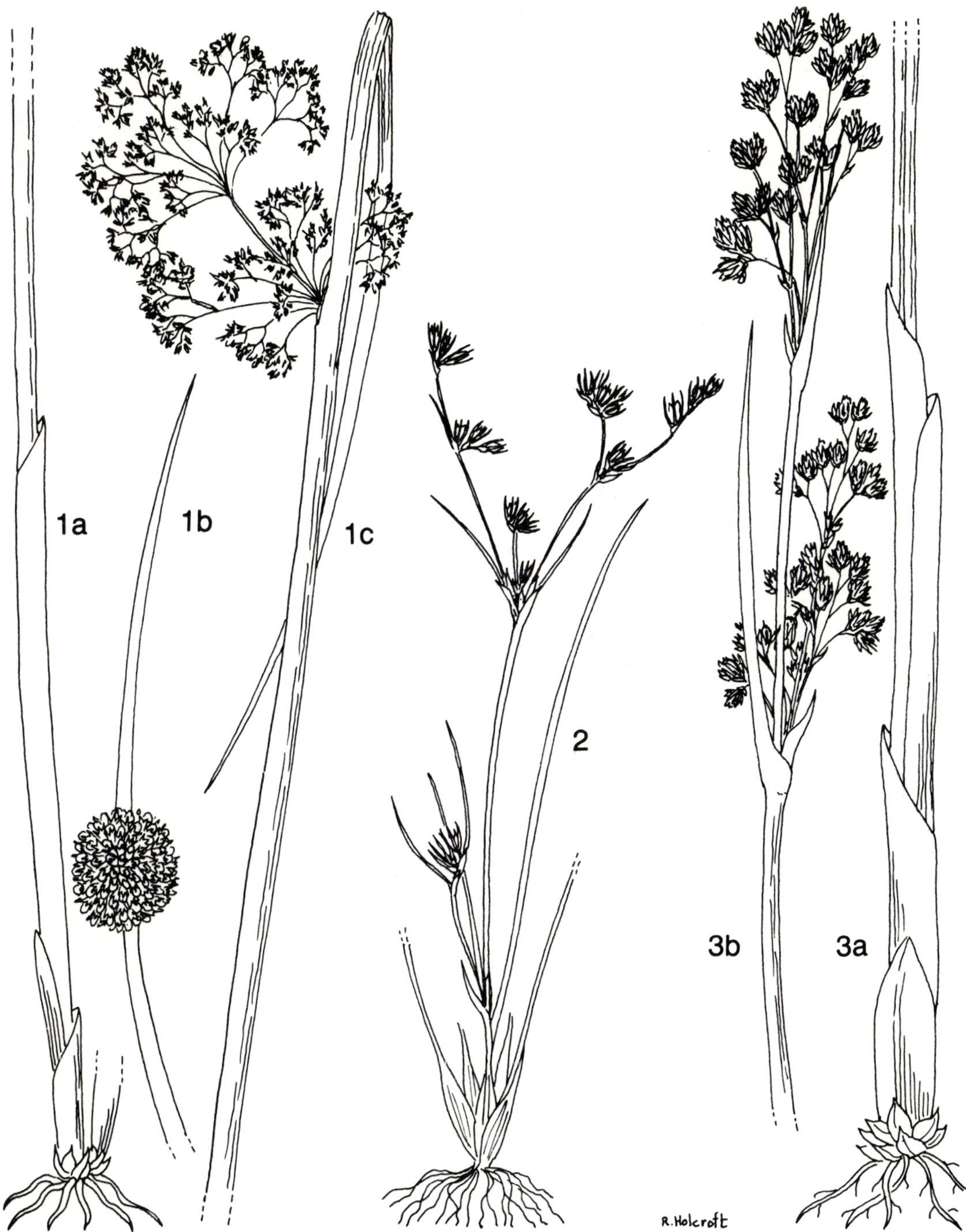

FIG. 23.—1, **Juncus effusus**: 1a, basal part of plant, × 0,5; 1b, inflorescence in bud, × 1; 1c, mature inflorescence, × 0,5 (*Behr* 156). 2, **J. bufonius**: habit, × 1 (*Hugo* 688). 3, **J. rigidus**: 3a, basal part of plant, × 1; 3b, inflorescence, × 0,5 (*Louw* 1145).

1. **Juncus bufonius** *L.*, Sp. Pl. 328 (1753); Buchen. in Abh. naturw. Ver. Bremen 4: 416 (1875), and in Pflanzenreich 4, 36 (Heft 25): 105 (1906); Bak. in F.C. 7: 23 (1897); Adamson in J. Linn. Soc., Bot. 50: 3 (1935); Carter in F.T.E.A. Juncaceae: 2 (1966); M. Friedrich et al. in F.S.W.A. 156: 2 (1967); Cutler in Anat. Monocot. 4: 27, etc. (1969). Type: Europe, LINN 449.24 (LINN, lecto.).

J. dinteri V. Poelln. in Feddes Reprium 48: 173 (1940). Type: South West Africa/Namibia, Garies, *Dinter* 4254 (B, holo.).

Annuals, tufted, soft, 30–300 mm tall. *Leaves* few, radical and cauline, narrowly linear-acuminate, with an orange, loosely sheathing base. *Flowers* in leafy panicles or lateral secund spikes, sessile, solitary; bracts 3, hyaline. *Tepals* c. 5 mm long, green with a hyaline border, outer aristate, inner shorter, acute. *Stamens* 6(–3), short; anthers about as long as filaments. *Ovary* 3-locular; style and stigmas about as long as perianth. *Capsule* included, ellipsoid, obtuse; seed obovoid, c. 0,4 mm long, base rounded, apex truncate, ribbed, pale brown, becoming mucilaginous when wet. Fig. 23: 2.

An innocuous introduced weed fairly frequent in disturbed places in the south-western Cape. Also recorded from the Orange Free State and Natal where it is rare. Cosmopolitan in temperate regions. Flowering in spring. Map 57.

Vouchers: *Acocks* 17429; *Esterhuysen* 12638; *Paterson* 16; *Potts* 2870; *Sim* in PRE 38245.

2. **Juncus tenuis** *Willd.*, Sp. Pl. 2: 214 (1799); Buchen. in Pflanzenreich 4, 36 (Heft 25): 115 (1906). Cutler in Anat. Monocot. 4: 25, etc. (1969); Hilliard & Burtt in Notes

R. bot. Gdn Edinb. 40, 2: 280 (1982). Type: America (B, Willd. Herb. 6888).

Perennials, tufted, up to 0,4 m tall; rhizomes compact; roots thin. *Leaves* narrowly linear-acuminate, grass-like, light green, sheathing bases auriculate. *Flowers* in exserted, compound, irregular dichasia, lowest bracts much longer than few-flowered, contracted, secund spikelets borne on branchlets of varying length; floral bracts small, acute. *Tepals* linear-acuminate, 3–4 mm long, aristate, green with hyaline margin. *Stamens* 6, short; anthers shorter than filaments. *Ovary* short; style and stigmas shorter than tepals. *Capsule* globose, 3,5 mm in diam.; seeds obliquely obovate, c. 0,35 mm long, light brown with short white apiculus, becoming mucilaginous when wetted.

A species described from America, now widespread in South America, Europe, Australia and elsewhere. Recently collected in the eastern Transvaal and the Natal Drakensberg at high altitudes. Map 57.

Vouchers: *Acocks* 22141; *Killick & Vahrmeijer* 3791; *Musil* 258, 267; *Werdermann & Oberdieck* 1448.

The minute seeds, which become mucilaginous and sticky when moistened, can be easily transported by migratory wading birds.

3. **Juncus imbricatus** *La Harpe*, Monogr. Joncées: 149 (1827). Type: South America, *Commerson* s.n. (G).

J. chamissonis Kunth, Enum. Pl. 3: 348 (1841); Adamson in S. Afr. J. Sci. 28: 251 (1931), and in J. Linn. Soc., Bot. 50: 4 (1935). *J. imbricatus* var. *chamissonis* (Kunth) Buchen. in Pflanzenreich 4, 36 (Heft 25): 122 (1906). Type: Chile, *Chamisso* s.n. (B†).

Sclerotic, caespitose perennials up to 0,6 m tall. *Rhizome* compact, fibrous, with hard, woody roots densely covered by root-hairs. *Leaves* basal, 1–2 per shoot, filiform, sulcate, erect, expanded at base into tight auriculate sheaths; cataphylls present. *Inflorescence* pseudolateral, many-flowered, lowest bract forming continuation of stem, short or exserted; floral bracts and bracteoles short. *Flowers* solitary, secundly arranged on few to many short, lateral rachides. *Tepals* ovate-acute, 4–5 mm long, shiny, subequal, hard. *Stamens* 6, short; anthers and filaments about equally long. *Ovary* ellipsoid; style 0; stigmas c. 1 mm long. *Capsule* narrowly ovoid-triangular, 5 mm long, truncate, exserted, shiny, placentas parietal, intruding, white, irregularly pectinate with seeds attached to fairly long funicles; seeds oblong-globose, 0,4 mm

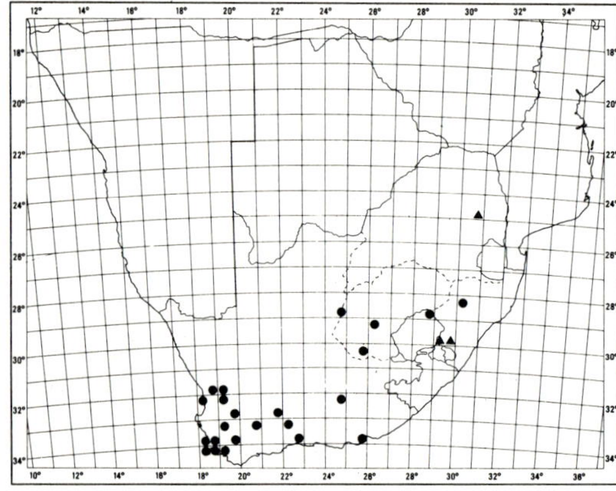

MAP 57.— ● **Juncus bufonius**
▲ **Juncus tenuis**

long, brown, apex and base with a membranous apiculus.

An introduction from South America, rarely collected around Cape Town; flowering in spring. Map 58.

Vouchers: *Adamson* 4, 7, 336; *Mauve* 5035.

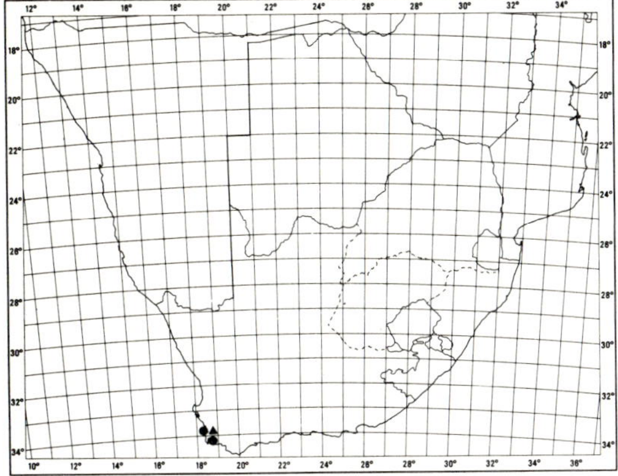

MAP 58.— ● **Juncus imbricatus**
▲ **Juncus capillaceus**

4. **Juncus capillaceus** *Lam.*, Encycl. 3: 266 (1789); Kunth, Enum. Pl. 3: 350 (1841); Buchen. in Bot. Jb. 12: 199 (1890), and in Pflanzenreich 4, 36 (Heft 25): 122 (1906); Adamson in S. Afr. J. Sci. 28: 251 (1931), and in J. Linn. Soc., Bot. 50: 5 (1935); Cutler in Anat. Monocot. 4: 41, etc. (1969). Type: South America, Monte Video, *Commerson* s.n. (P, holo.).

Perennials up to 0,3 m tall. *Rhizome* horizontal, compressed; roots woody, covered by a thick layer of root-hairs. *Shoots* close together, fibrous at base. *Leaves* basal, 1–3 per shoot, filiform, erect, ribbed, longer than inflorescence, expanded below into tight auriculate sheaths; cataphylls present. *Inflorescence* pseudo-lateral, few-flowered, lowest bract long, forming continuation of stem; floral bracts and bracteoles small; flowers 2–10, solitary, secund, clustered. *Tepals* ovate, acute, c. 2–3 mm long, subequal. *Stamens* 6, very short; anthers about as long as filaments. *Ovary* ovoid, obtuse. *Capsule* ovoid, 3 mm long, obtuse to acute, enclosed in perianth; parietal placentas forming a smooth swollen ridge; funicles short; seeds ellipsoid, 0,5 mm long, brown, apex and base white-apiculate, connected by a white ridge.

An introduction from South America, collected a few times by Adamson around Cape Town in disturbed areas. Flowering and fruiting March–April. Map 58.

Vouchers: *Adamson* 1, 195; *Mauve & Fairall* 5040.

5. **Juncus effusus** *L.*, Sp. Pl. 326 (1753); Bak. in F.C. 7: 18 (1897); Buchen. in Pflanzenreich 4, 36 (Heft 25): 115 (1906); Adamson in J. Linn. Soc., Bot. 50: 5 (1935); Weim. in Svensk bot. Tidskr. 40: 143 (1946); Cutler in Anat. Monocot. 4: 25, etc. (1969); Carter in F.T.E.A. Juncaceae: 2 (1966). Type: Europe (LINN 449.3, lecto.).

Perennials, tufted, up to 1 m tall, variable. *Rhizomes* horizontal, matted; roots many, fairly thick. *Leaves* basal, cauliform. *Stems* naked, erect, terete, 2–3 mm in diam., smooth, bright green, pith asterisciform, continuous; cataphylls 70–170 mm long, sheathing stems and leaves. *Inflorescence* pseudolateral, subtending leaf forming continuation of stem; flowers very numerous and very small, in globose to lax, decompound anthelae; bracts 3 per flower, small, hyaline. *Tepals* acute, c. 2–3 mm long, green, subsimilar. *Stamens* 3 (rarely 6); filaments about as long as anthers. *Ovary* obtuse; style 0; stigmas short. *Capsule* ovoid to rounded, retuse, membranous, pale yellowish; seeds 0,4 mm long, fusiform but bulging on one side, golden brown, reticulate with strong wavy lines, becoming mucilaginous when wet. Fig. 23: 1.

Cosmopolitan but more common in the northern hemisphere. Widespread in Africa; not common in Southern Africa. Recorded from Transvaal, Natal, Lesotho and the southern and south-western Cape; in swamps or streambeds. Flowering in spring and summer. Map 59.

Vouchers: *Coetzee* 839; *Devenish* 280; *Lubke* 284; *Mogg* 877; *Moll* 1457.

6. **Juncus inflexus** *L.*, Sp. Pl. 326 (1753); Butcher, New Ill. Brit. Flora 2: 682, t. 1478 (1961); Cutler in Anat. Monocot. 4: 30, etc.

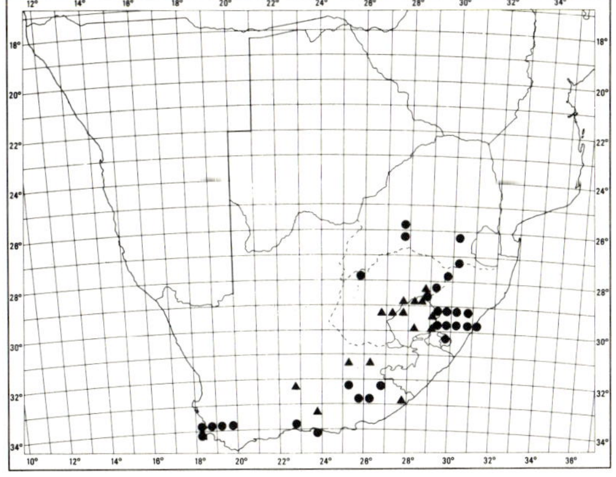

MAP 59.— ● **Juncus effusus**
▲ **Juncus inflexus**

(1969); Snogerup in Fl. Iran. Juncaceae: 8 (1971). Type: southern Europe.

J. glaucus Ehrh. var. *acutissimus* Buchen. in Abh. naturw. Ver. Bremen 4: 417, t. 5 (as 4) (1875), and in Natürl. PflFam. 4, 36: 133 (1906); Bak. in F.C. 7: 18 (1897). *J. acutissimus* (Buchen.) Adamson in J. Linn. Soc., Bot. 50: 6 (1935). Type: Cape, Wodehouse, Klein Buffels Vallei near Gaatjie, *Drège* 8796c (W, holo.!).

Perennial, tufted; rhizome matted, horizontal; roots thick. *Stems* up to 1 m tall, leafless, glaucous, wiry, sheathed basally with chestnut-brown, shiny cataphylls; pith interrupted. *Inflorescence* pseudolateral, subtending leaf forming apparent continuation of stem, about ¼ of length of stem; anthelae compact, many-flowered, mostly on short to very short branchlets. *Tepals* linear, acute, outer somewhat longer, c. 4 mm long, firm, ribbed. *Stamens* 6, c. 2 mm long; filaments filiform, equalling anthers. *Ovary* cylindrical; style present with 3 long stigmas. *Capsule* ellipsoid, tricostate, 0,3 mm long, hard, dark brown, shiny, apiculate; seeds 0,4 mm long, ovoid, bulging unilaterally, reticulate, golden yellow, with a dark, obtuse apex and base, tips transparent.

Widespread in Europe, Africa, Asia; introduced in New Zealand; in Southern Africa predominantly along the eastern escarpment, usually at high altitudes: Orange Free State, Lesotho and the eastern Cape; locally frequent along streambanks and swamps. Flowering in spring. Map 59.

Vouchers: *Acocks* 18660; *Flanagan* 1666; *Liebenberg* 5834, 7018; *Lubke* 221.

7. **Juncus acutus** *L.*, Sp. Pl. 325 (1753).

subsp. **leopoldii** *(Parl.) Snogerup* in Bot. Notiser 131: 187 (1978). Type: Cape, *Ecklon & Zeyher* 4308 (FI, holo., fide Snogerup in litt.).

J. leopoldii Parl. in G. bot. ital. 2: 324 (1846). *J. acutus* L. var. *leopoldii* (Parl.) Buchen. in Abh. naturw. Ver. Bremen 4: 421, t. 5 (1875); Bak. in F.C. 7: 19 (1897).

J. acutus sensu Adamson in J. Linn. Soc., Bot. 50: 7 (1935); Cutler in Anat. Monocot. 4: 20, etc. (1969).

Perennial, tufted, usually tall and hard. *Rhizomes* horizontal, compressed; woody roots covered with a thick velamen of roothairs. *Stems* 1–2 m tall, coarse, terete, smooth. *Leaf* one per shoot, forming a cylindric pungent blade as tall as inflorescence, below enveloped by 2–3 loose, brown, sheathing cataphylls. *Flowers* in pseudolateral, decompound panicles, subtending leaf forming continuation of stem; floral glomerules on

numerous ascending branches and branchlets of varying length. *Tepals* c. 3 mm long, canaliculate, outer somewhat shorter, apiculate, inner with membranous auricles extended above retuse mucronate apex. *Stamens* 6; anthers large, 1,5 mm long, red; filaments short, flat. *Ovary* ovoid; style short; stigmas swollen, c. 1,5 mm long, red, just exserted. *Capsule* exserted, globose, c. 4,5 mm diam., apiculate, shiny, reddish brown; seed allantoid, c. 1 mm long, with a white membranous appendage on either side.

Widespread in temperate zones of the Western Hemisphere. Confined to the Cape Province where it is known from the north, east and south-west but not yet from the Peninsula. Halophytic, forming large colonies. Flowering in spring. Map 60.

Vouchers: *Acocks* 16339; *Liebenberg* 6696; *M. Schlechter* 20; *Sim* 162; *Van Breda* 58.

8. **Juncus kraussii** *Hochst.* in Flora 28: 342 (1845); Buchen. in Abh. naturw. Ver. Bremen 4: 418, t. 4 (1875), and in Bot. Jb.

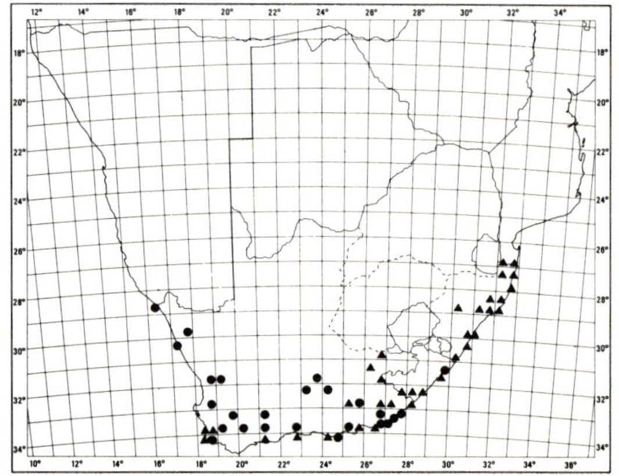

MAP 60.— ● **Juncus acutus** subsp. **leopoldii**
▲ **Juncus kraussii**

12: 255 (1890), and in Pflanzenreich 4, 36 (Heft 25): 153 (1906); Adamson in J. Linn. Soc., Bot. 50: 8 (1935) including vars *effusus* Adamson and *parviflorus* Adamson. Type: Cape, Tsitsikamma, *Krauss* s.n. (G!; W, iso.!).

J. spretus Schultes & Schultes f. in Roem. & Schult., Syst. Veg. 7, 2: 1655 (1830), non rite publ.; nom. prov. Type: Cape, *Ecklon* 903.

J. maritimus sensu auct., non Lam., p. p.; Bak. in F.C. 7: 19 (1897).

J. fasciculiflorus Adamson in J. Linn. Soc., Bot. 50: 8 (1935). Type: Cape, Ceres, Swartruggens by Waterfall, *Adamson* 123 (BOL, holo.!).

Perennials, tufted, hard, dark green, c. 1 m or more tall. *Rhizomes* compact, horizontal; thick roots covered with velamen of root-hairs. *Stems* erect, terete, hard. *Leaves* cauliform, terete, pungent, few per shoot, about as long as stems, sheaths auriculate; cataphylls shiny. *Inflorescence* pseudolateral, the subtending bract forming a continuation of the stem; plumose, decompound with the many branches and branchlets of different lengths. *Flowers* in compact clusters of 3–6. *Tepals* linear-acuminate, c. 3 mm long, hard with age. *Stamens* 6, anthers c. 2 mm long with short filaments. *Ovary* ovoid; style short; stigmas exserted. *Capsule* narrowly ellipsoid, 3-angled, apiculate, woody, shiny brown, enclosed in perianth or slightly longer; seeds ellipsoid with one side bulging, 0,6 mm long, funicle and apex pointed, short, with a white membranous connecting ridge, reticulate, brown.

A widespread south circumpolar species also occurring in Australia and South America. Widespread in Southern Africa, from the Cape Peninsula eastwards along the coast to Natal and Mozambique. The furthest inland record appears to be from Dalton Bridge, Bushmans River in Natal. Forming large colonies, halophytic. Flowering in early summer. Map 60.

Vouchers: *Acocks* 10657; *Adamson* 122; *Flanagan* 3437; *Musil* 147; *Taylor* 6667; *Ward* 616.

9. Juncus rigidus *Desf.*, Fl. Atlant. 1: 312 (1800); Snogerup in Fl. Iran. Juncaceae: 4, t. 1, f. 1 (1971). Type: Africa boreo-occidentalis, collector unknown (P, in Herb. Fl. Atlant., teste Snogerup).

J. caffer Bertol. in Memorie R. Accad. Sci. Ist. Bologna, sér. 3: 253 (1851). Type: Mozambique, Inhambane, *Fornasinio* s.n. (BOLO, holo.).

J. maritimus Lam. var. *arabicus* Aschers. & Buchen. in Boiss., Fl. Or. 5: 354 (1882). *J. arabicus* (Aschers. & Buchen.) Adamson in J. Linn. Soc., Bot. 50: 10 (1935); M. Friedrich et al. in F.S.W.A. 156: 2 (1967). Type: Egypt, *Hausknecht* (BM; K; G-Boiss.).

J. maritimus sensu Bak. in F.C. 7: 19 (1897) and in F.T.A. 8: 93 (1901), non Lam.

Perennial, tufted, hard, with coarse, compact horizontal rhizomes and thick roots covered with a velamen of root hairs. *Stems* up to 1,5 m tall, terete, hard. *Leaves* terete, pungent, as long as stems, sheaths auriculate; cataphylls present. *Inflorescence* pseudolateral with subtending bract forming continuation of stem, decompound, interrupted, many-flowered. *Flowers* usually solitary or 2–few together on branchlets of varying lengths. *Tepals* linear-acuminate, with membranous margins, 4–5 mm long, usually straw-coloured. *Stamens* 6; filaments short; anthers c. 1 mm long. *Ovary* ovoid; style short; stigmas about as long as style, usually exserted sideways. *Capsule* narrowly cylindrical, tricostate, acute, light brown, exserted from perianth; seeds somewhat fusiform with funicle and apex forming long, white appendages. Fig. 23: 3.

Recorded from the Mediterranean Region and as far east as Pakistan; a widespread halophyte in Africa; in Southern Africa common in the interior, from South West Africa/Namibia to Transvaal, Orange Free State and Cape; forming large colonies. Map 61.

Vouchers: *Acocks* 9702, 10847; *Merxmüller & Giess* 30654; *Van Zinderen Bakker* 94; *Verdoorn* 1024; *Wilman in TRV* 19814.

Closely related to *J. kraussii* (above) but the flowers more loosely arranged, the narrow capsule distinctly exserted and the seeds with long appendages on either side.

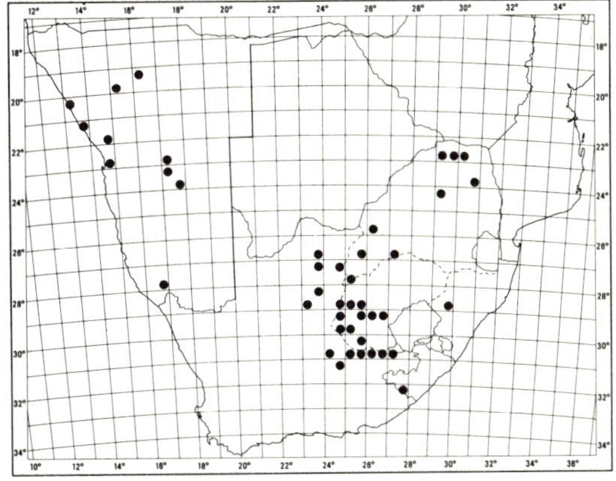

MAP 61.— **Juncus rigidus**

10. Juncus punctorius *L. f.*, Suppl. 208 (1782); Thunb., Prodr. 1: 66 (1794); Buchen. in Abh. naturw. Ver. Bremen 4: 424, t. 8 (1875) and in Natürl. PflFam. 4, 36: 163 (1906), includ. var. *exaltatus* (Decne.) Buchen., l.c. 129, t. 8; Bak. in F.C. 7: 20 (1897); Adamson in J. Linn. Soc., Bot. 50: 11 (1935); M. Friedrich et al. in F.S.W.A. 156: 3 (1967). Type: Cape Peninsula, *Thunberg & Sonnerat* (LINN 449.15, holo., PRE, photo.!).

Perennial, tufted, robust, hard, pale green, with horizontal woody rhizomes bearing numerous thin roots. *Flowering stem* up to 1,5 m tall, with 2–3 basal sheaths and 1 leaf arising near middle, septate with central and lateral lacunae, basal sheath obtusely auriculate. *Basal leaves* cauliform ('sterile stems'), present beside flowering stems. *Flowers* in pseudolateral, decompound, compact to somewhat effuse panicles with up to 100 globose, compact capitula. *Tepals* 2,5 mm long, subequal, outer acute. *Stamens* with anthers about as long as filaments. *Ovary* ovoid; style and stigmas fairly long. *Capsule* ovoid, trigonous, 2,5 mm long, apiculate; seeds ovoid-globose, obtuse, with large reticulae. Fig. 24: 2.

Distribution disjunct: in South West Africa/Namibia, Transvaal, Orange Free State, Lesotho, Natal and Cape. Also known from north-eastern Africa, Algiers, Arabia and Beluchistan. A hygrophyte, forming large colonies. Flowering in summer. Map 62.

Vouchers: *Adamson* 14; *Dieterlen* 1314; *Gerstner* 4345; *Killick* 1434; *Roberts* 2810.

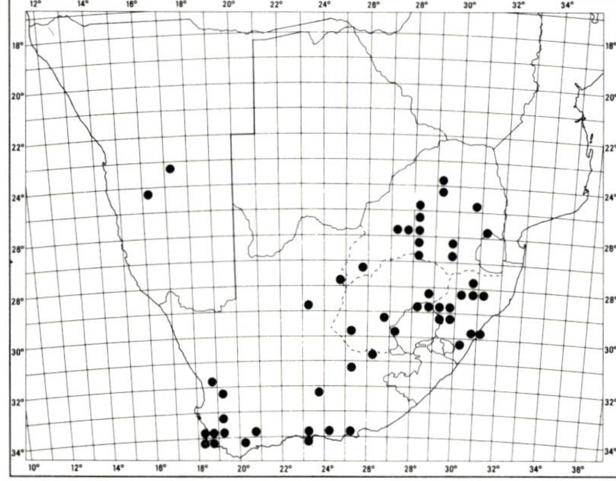

MAP 62.— **Juncus punctorius**

11. Juncus oxycarpus *Kunth*, Enum. Pl. 3: 336 (1841); Buchen. in Abh. naturw. Ver. Bremen 4: 431 (1875) and in Pflanzenreich 4, 36 (Heft 25): 196 (1906); Bak. in F.C. 7: 20 (1897), and in F.T.A. 8: 93 (1901); Adamson in J. Linn. Soc., Bot. 50: 12 (1935); Weim. in Svensk bot. Tidskr. 40: 166 (1946) as var. *australis* Weim.; Carter in F.T.E.A. Juncaceae: 3 (1966). Syntypes: Cape, Liesbeek River, *Bergius* (B†); Berg River near Paarl, *Drège* s.n. (K, iso.!).

J. brevistilus Buchen. in Abh. natuw. Ver. Bremen 4: 433, t. 8 (1875) and in Pflanzenreich 4, 36 (Heft 25): 200 (1906); Bak. in F.C. 7: 20 (1895); Adamson in J. Linn. Soc., Bot. 50: 14 (1935), as *brevistylus*. Type: Cape or Natal, locality and collector unknown.

J. gentilis N.E. Br. in Kew Bull. 2: 83 (1914). Type: Transvaal, near Modderfontein, *Conrath* 1173 (K, holo.!).

J. oxycarpus var. *microcephalus* Adamson, l.c. 13. Syntypes: Cape, Riversdale, *Muir* 3385 (BOL!); Grahamstown, *Dyer* 173 (BOL!), etc.

J. suboxycarpus Adamson, l.c. 14. Type: Natal, Clairmont, *Schlechter* 3043 (BOL; holo.!; PRE!).

Perennial, tufted; rhizome caespitose with many thin, branched roots. *Stems* erect, 0,4–0,8 m tall, occasionally stoloniferous when in running water, dwarfed at high altitudes. *Leaves* 3–5, basal and cauline, terete, distinctly septate, apex pointed, base with open, auriculate sheaths; cataphylls 1–4, acute. *Inflorescence* exserted, decompound, divaricate, capitula globose or semiglobose, many-flowered, on wiry branchlets of varying lengths; occasionally flowers changing into vegetative buds; bracts broadly ovate, aristate, membranous. *Tepals* subequal, linear-acuminate, 3–4 mm long, aristate. *Stamens* 3 or 6; anthers shorter than filaments. *Ovary* 1-celled, ovate-acute; style short, with 3 longer red stigmas. *Capsule* unilocular, oblong-ovoid, obtuse, apiculate, trigonous, as long as perianth; seeds ovoid, 0,7 mm long; testa coarsely reticulate, with raised bars forming reticulate pattern. Fig. 24: 1.

Widespread but confined to Africa; usually in very wet areas on streambanks and in swamps. Recorded from Southern Africa to tropical eastern Africa as far as Eritrea. See also notes under *J. exsertus* (below). Map 63.

Vouchers: *Acocks* 20435; *Compton* 31285; *Killick* 1087; *Leistner* 1376; *Morwe* 45; *Pott* 5160.

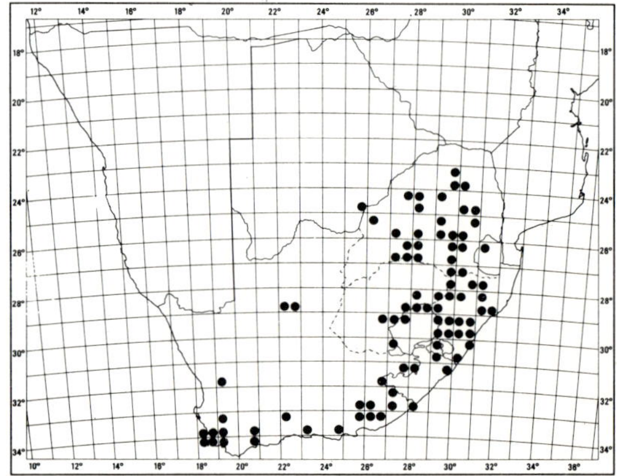

MAP 63.— **Juncus oxycarpus**

12. Juncus exsertus *Buchen.* in Abh. naturw. Ver. Bremen 4: 435, t. 5 (1875), and in Pflanzenreich 4, 36: 196 (1906); Bak. in F.C. 7: 21 (1897); Adamson in J. Linn. Soc., Bot. 50: 15 (1935); M. Friedrich et al. in F.S.W.A. 156: 2 (1967). Syntypes: Cape, Swartkops River, *Zeyher* 103 (B†; BOL!); Worcester, at the Waterfall, *Ecklon & Zeyher*, pro parte

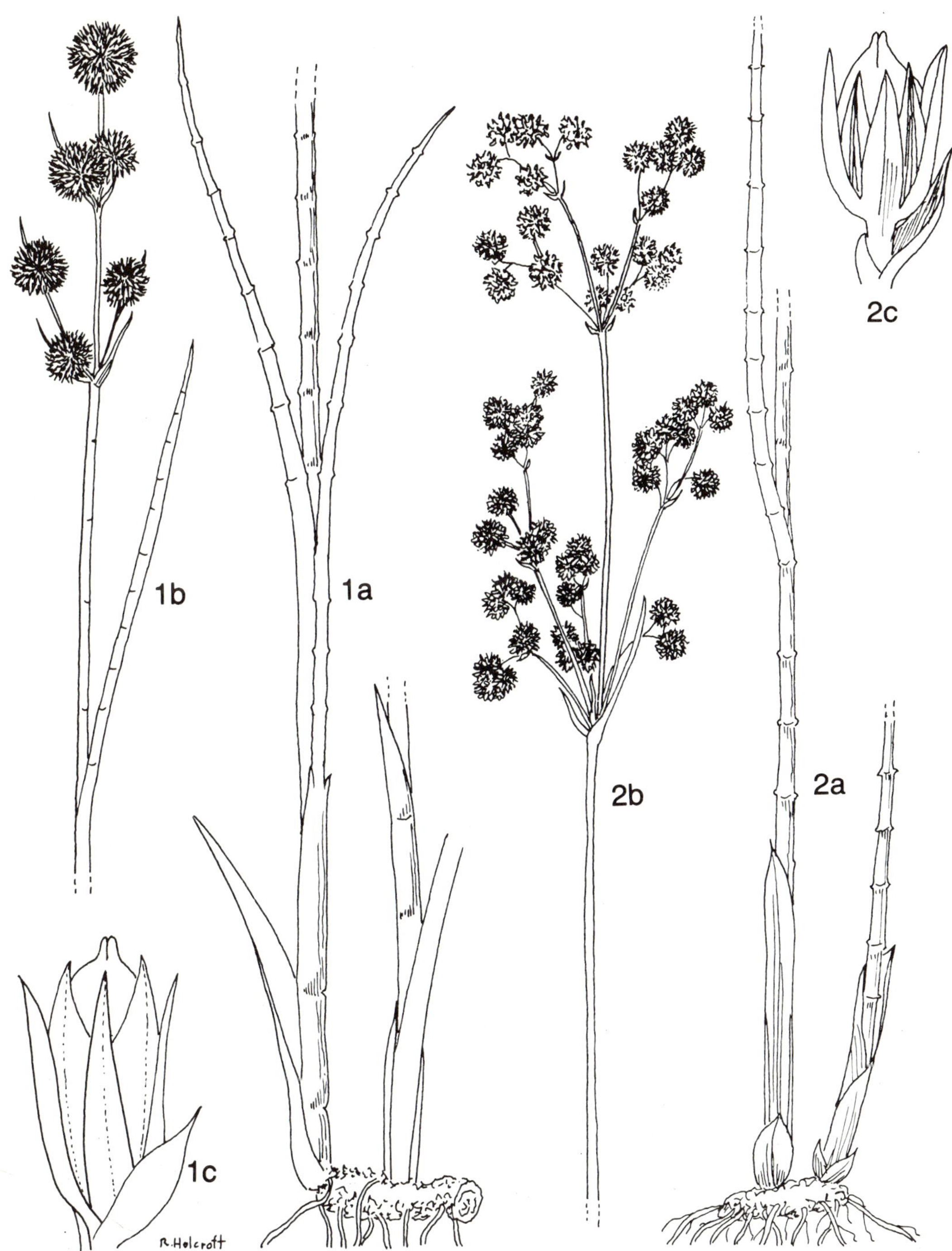

FIG. 24.—1, **Juncus oxycarpus**: 1a, part of plant, × 1; 1b, part of inflorescence, × 0,5; 1c, flower in fruit, × 0,1 (*Dieterlen* 767b). 2, **J. punctorius**: 2a, basal part of plant, × 0,5; 2b, inflorescence, × 0,5; 2c, fruit surrounded by tepals, × 10 (*Leendertz* 1652).

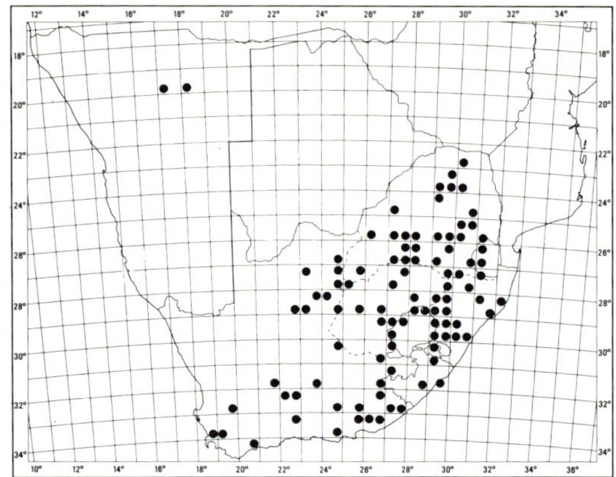

MAP 64.— **Juncus exsertus**

(B†; PRE!); Sundays River at Graaff-Reinet, *Bolus* 188 (BOL!).

J. rostratus Buchen. in Abh. naturw. Ver. Bremen 4: 437, t. 5 (1875), and in Pflanzenreich 4, 36: 196 (1906); Bak. in F.C. 7: 21 (1897); Adamson in J. Linn. Soc., Bot. 50: 16 (1935). Syntypes: Cape, Swartkops River, *Ecklon & Zeyher* s.n. (S); Bashee River, *Drège* 4465 (K!; G!).

Similar to *J. oxycarpus* (above) but capitula are usually fewer-flowered, obconic; capsules thin, slender, well exserted from perianth; raised bars forming reticulate pattern on testa of seeds (× 1 000 enlargement) moniliform, not smooth. Occasionally flowers turn into vegetative buds which may form new plants.

Confined to Southern Africa and Zimbabwe. Widespread in South West Africa/Namibia, Transvaal, Swaziland, Orange Free State, Lesotho, Natal and Cape; on riverbanks or in marshes. Map 64.

Vouchers: *Adamson* 185; *Arnold* 414; *Compton* 27415; *Lubke* 286; *Moll* 663; *Werger* 299.

Weimarck in Svensk bot. Tidskr. 40: 166 (1946) considered *J. exsertus*, *J. brevistylus*, *J. rostratus* and *J. suboxycarpus* to be synonyms of *J. oxycarpus*. While *J. brevistylus* and *J. suboxycarpus* cannot be distinguished from the now ample material of *J. oxycarpus*, *J. exsertus* can usually be easily recognized. *J. rostratus*, with its exserted capsule, is a synonym of *J. exsertus*.

13. **Juncus lomatophyllus** Spreng., Neue Entdeck. 2: 108 (1821); Buchen. in Abh. naturw. Ver. Bremen 4: 466 (1875), and in Pflanzenreich 4, 36 (Heft 25): 247 (1906); Bak. in F.C. 7: 27 (1897); Adamson in J. Linn. Soc., Bot. 50: 18 (1935). Type: Cape Peninsula, *Bergius* (B†).

J. cymosus Lam., Encycl. 3: 267 (1789), nom. rej. Type: Cape, *Sonnerat* (P, holo., p. p., for inflorescence; leaf excluded).

J. cephalotes sensu Thunb., Prodr. 66 (1794), Fl. Cap. edn 2, 337 (1823), p. p. Type: Cape, *Thunberg* sheet 8709, l.h. specimen; sheet 8711.

J. capensis var. *latifolius* E. Mey. in Syn. Junc. 48 (1822). Type: Cape, Simonsberg, *Drège* aa.

J. lomatophyllus var. *aristatus* Buchen., l.c. Type: Cape, Du Toitskloof, *Drège* f.

J. lomatophyllus var. *lutescens* Buchen., l.c. Type: Cape, Du Toitskloof, *Drège* a.

J. viridifolius Adamson in J. Linn. Soc., Bot. 50: 20 (1935). Type: Cape, Table Mountain, *Adamson* 247 (BOL, holo.!).

J. lomatophyllus var. *congestus* Adamson, l.c. Type: Cape, Swartberg Pass, *Adamson* 181 (BOL, lecto.!).

Perennials with spreading to decumbent, densely leafy stolons; roots thin. *Leaves* subrosulate to spirally arranged, broadly linear, up to 200 mm long and 10–15 mm broad, flat, soft, apex pointed, base often reddish, tubular when very young, soon splitting, margin narrowly membranous. *Inflorescence* up to 0,75 m tall, a decompound dichasium but much reduced in montane habitats at high altitudes, subtended by a short leaf-like basal bract. *Capitula* semi-globose with 6–12 flowers. *Tepals* dark brown, outer 5 mm long, long-aristate, inner shorter, acute, margins membranous. *Stamens* 6; anthers 1,5 mm long, on very short filaments. *Ovary* 3-locular, ovoid, attenuated into a long, red style and 3 long, red, exserted stigmas 2 mm long. *Capsule* ovoid, rostrate, 2,5 mm long; seeds oblong-globose, 0,8 mm long, pointed on both sides, epidermis membranous, hygroscopic. Fig. 25: 4.

Widespread along the eastern escarpment in Transvaal, Swaziland, Lesotho, Natal and Cape; in the south-western Cape from the Peninsula to Clanwilliam; also in Zimbabwe and recorded as an adventive in New Zealand.

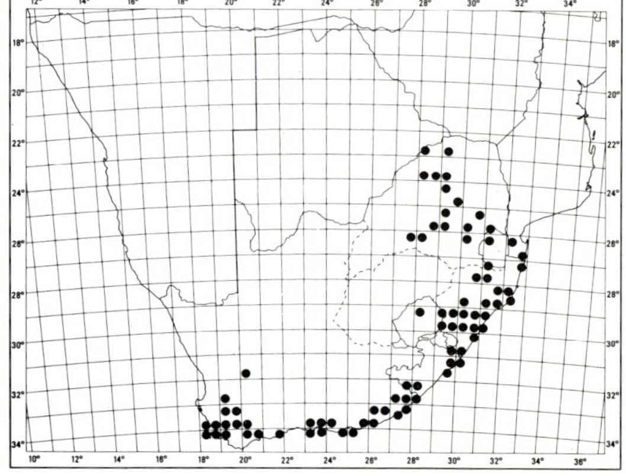

MAP 65.— **Juncus lomatophyllus**

Beside streams or in permanently wet, montane habitats. Flowering in late spring and summer. Map 65.

Vouchers: *Codd & Dyer* 9010; *Compton* 26351; *Hanekom* 1199; *Muller* 2112; *Strey* 9279.

A number of collections from the eastern Cape appear to be intermediate between *J. lomatophyllus* and *J. capensis* (below). Examples are *Flanagan* 1006 from near Komgha (3227DB) and *Galpin* 7682 from the Keiskamma River mouth (3327AD). These putative hybrids appear to be more upright in habit, like *J. capensis*, but the leaves are somewhat broader. The anthers are small and possibly sterile. The inflorescences resemble those of *J. lomatophyllus*. See also notes under *J. capensis* (below).

14. **Juncus capensis** *Thunb.*, Prodr. 1: 66 (1794); Buchen. in Abh. naturw. Ver. Bremen 4: 482 (1875); Bak. in F.C. 7: 26 (1897); Adamson in J. Linn. Soc., Bot. 50: 27 (1935); Cutler in Anat. Monocot. 4: 47, 72 (1969). Type: Cape, *Thunberg* 8702 (UPS, lecto.).

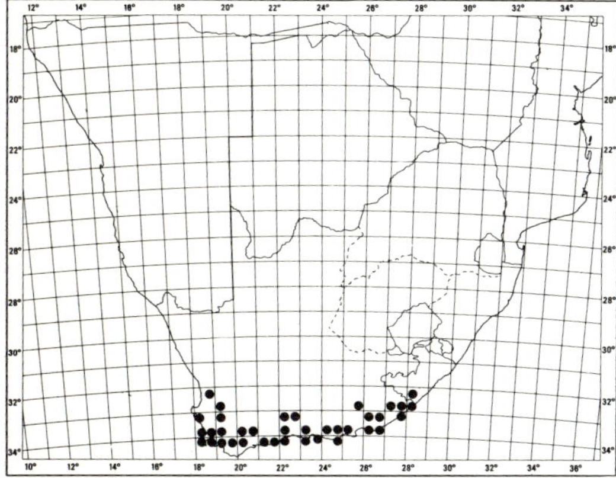

MAP 66.— **Juncus capensis**

J. sulcatus Hochst. in Krauss in Flora 28: 342 (1845). *J. capensis* subsp. *angustifolius* var. *flaccidus* forma *depauperata* Buchen., l.c. 489. Type: Cape, Tsitsikamma, *Krauss* s.n. (W, holo.!, PRE, photo.!; G!).

J. anonymus Steud., Syn. Pl. Glum. 304 (1855); Buchen., l.c. 478; Adamson, l.c. 26 (1935). Type: Cape, Du Toitskloof, *Drège* 1604a (G!; K!; W, iso.!, PRE, photo.!).

J. delicatulus Steud., Syn. Pl. Glum. 2: 304 (1855). *J. capensis* subsp. *delicatulus* (Steud.) Buchen., l.c. 490. *J. capensis* var. *delicatulus* (Steud.) Adamson, l.c. 28. Type: Cape, Grahamstown valley, *Drège* 1604e (W!; G!, iso., PRE, photo.!).

J. flaccidus Steud., Syn. Pl. Glum. 2: 303 (1855). *J. capensis* subsp. *angustifolius* var. *flaccidus* Buchen., l.c. 488. *J. capensis* var. *ecklonii* Buchen. emend. Adamson, l.c. 27 (1935), p. p. Type: Cape, Table Mountain, *Bergius* (W!; PRE, photo.!).

J. indescriptus Steud., Syn. Pl. Glum. 2: 304 (1855); Buchen., l.c. 479; Adamson, l.c. 24 (1935). Type: Cape, Berg River near Paarl, *Drège* 1604h (G!; K!; W, iso.!, PRE, photo.!).

J. singularis Steud., Syn. Pl. Glum. 2: 302 (1855); Buchen., l.c. 438 (1875); Adamson, l.c. 30 (1935). Type: Cape, between Vanstadensberg and Bethelsdorp, *Drège* 1604b, pro parte (G!; W, iso.!, PRE, photo.!).

J. stenophyllus Steud., Syn. Pl. Glum. 2: 203 (1855); *J. capensis* subsp. *angustifolius* E. Mey. ex Buchen., p. p., l.c. 484, 485. Type: Cape, *Ecklon* 897, Unio itin. No. 35 (W, iso.!). Steudel cited no type but refers to *J. capensis angustifolius* E. Mey. Placed under var. *ecklonii* by Buchenau and Adamson.

J. acutangulus Buchen., l.c. 480 (1975); Bak. in F.C. 7: 27 (1897). Syntypes: Cape, Somerset West, moist places on the Cape flats, *Ecklon* 4318 (BOL!; W!); Wynberg, *Ecklon* 100 (W!, PRE, photo.!).

J. capensis subsp. *angustifolius* E. Mey. ex Buchen., p. p., l.c. 484. Type not designated.

J. capensis subsp. *angustifolius* var. *ecklonii* Buchen., l.c. 485. *J. capensis* var. *ecklonii* Buchen. emend. Adamson, l.c. 27, p. p. Type: Cape, Devil's Peak at foot, *Ecklon* 897, Unio itin. No. 35 (W, iso.!).

J. capensis subsp. *angustifolius* var. *sphagnetorum* Buchen., l.c. 489. *J. sphagnetorum* Adamson, l.c. 29. Type: Cape, Du Toitskloof, *Drège* 'J. cap. var. ang., cc' (W, holo.!).

J. capensis subsp. *angustifolius* var. *sphagnetorum* forma *frondescens* Buchen., l.c. 490. Type: Cape, Table Mountain, *Drège* 'J. cap. Thunb. var. angustifolius E. Mey. aa' (W, holo.!).

J. capensis subsp. *geniculatus* Buchen., l.c. 492. *J. apiculatus* Adamson, l.c. 27, p. p. Syntypes: Cape, Howieson's Poort, *MacOwan* 2019, 2020 (W!; BOL!; SAM!).

J. capensis subsp. *longifolius* E. Mey. ex Buchen., l.c. 482. Type not designated.

J. capensis subsp. *longifolius* var. *gracilior* Buchen., l.c. 483. *J. capensis* var. *ecklonii* Buchen. emend. Adamson, l.c. 27, pro parte. Type: Cape, *Bergius* (W, holo.!).

J. capensis subsp. *longifolius* var. *strictissimus* Buchen., l.c. 482. Type: Cape, Hottentottsholland, *Gueinzius* s.n. (W, holo!, PRE, photo.!).

J. capensis subsp. *parviflorus* Buchen., l.c. 491. Type: Cape, Swellendam, Rivier Zondereinde, *Krauss* s.n. (W, holo.!, PRE, photo.!).

J. apiculatus Adamson, l.c. 27 (1935). *J. capensis* subsp. *geniculatus* Buchen., l.c. 492, t. 11 (1875), pro parte. Syntypes: Cape, Howieson's Poort, *MacOwan* 2020 (BOL!; SAM!; W!); Transkei, Kentani, *Pegler* 1107 (BOL!).

J. atropurpureus Adamson, l.c. 30 (1935). *J. capensis* subsp. *longifolius* var. *strictissimus* Buchen., l.c. 482 (1875) pro parte. Type: Cape Flats near Lakeside, *Adamson* 171 (BOL, holo.!).

J. capensis var. *macranthus* Adamson, l.c. 28. Syntypes: Cape, Kommetjie, *Adamson* 88 and 89 (BOL!).

J. umbellatus Adamson, l.c. 38 (1935). Type: Cape, Berg River at Wellington, *Adamson* 313 (BOL, holo.!).

Tufted erect perennials with compact rhizomes, 60–600 mm tall; roots thin. *Leaves* numerous, grass-like, filiform, setaceous to narrowly linear, flat or inrolled, margin narrowly membranous below. *Inflorescence* usually exserted, very varied, a decompound dichasium to much reduced, subtended by a short bract; floral bracts tepaloid. *Capitula* semi-globose to globose, with 10–15–20 flowers. *Tepals* with outer ovate-aristate, 4–5 mm long, with dark brown stripes, becoming pale with age; inner shorter and broader, acute with membranous margins. *Stamens* 6; anthers long and narrow, often diverging at base; filaments short, flat. *Ovary* ovoid, attenuated into a red style 1 mm long, with 3 red exserted stigmas, c. 2 mm long. *Capsule* ovoid to narrowly oblong-ovoid, apiculate to rostrate, 3-locular; seed dark, oblong-globose, 0,5 mm long, obtuse, with a membranous covering, which becomes mucilaginous when wet. Fig. 25:1.

Widespread and common in the south-western Cape, extending east as far as the Transkei. Introduced in Australia and New Zealand. Map 66.

Vouchers: *Adamson* 237, 303; *Barker* 2853; *Esterhuysen* 4203; *Schlechter* 1935; *Zeyher* 443.

When Buchenau compiled his monograph on the Cape species in 1875, he mentioned on p. 462 that more ample collections might produce intermediate forms linking *J. acutangulus*, *J. anonymus* and *J. indescriptus* with *J. capensis*. He realized it was a polymorphic species. I was unable to divide the species into subspecies or varieties. These subdivisions mostly represent ecological growth forms and were interpreted as such by Buchenau himself. *J. umbellatus* Adamson represents an immature stage of *J. capensis*. These small plants match the younger paedogenic offspring found amongst the tufts of mature plants.

In the herbarium difficulty was experienced in separating *J. capensis* from certain narrow-leaved forms of *J. lomatophyllus*. The flowers, capsules and seeds appear to be similar but while normally the creeping, broad-leaved stolons characterize *J. lomatophyllus*, some forms of *J. capensis* with somewhat broader leaves come close to narrow-leaved forms of *J. lomatophyllus*. This will need research in situ; see note under *J. lomatophyllus* (no. 13).

15. Juncus dregeanus *Kunth*, Enum. Pl. 3: 344 (1841); Buchen. in Abh. naturw. Ver. Bremen 4: 462, t. 9 (1875), and in Pflanzenreich 4, 36: 251 (1906); Bak. in F.C. 7: 25 (1897); Adamson in J. Linn. Soc., Bot. 50: 21 (1935). Type: between the Cape Colony and Port Natal, *Drège* 4387 (B†).

J. cephalotes sensu Thunb., Prodr. 66 (1794), p. p., for r.h. specimen on sheet 8709 (UPS).

J. submonocephalus Steud., Syn. Pl. Glum. 2: 303 (1855). *J. dregeanus* var. *submonocephalus* (Steud.) Buchen., l.c. 463. Type: Cape, *Drège* 1604f (W!, dissected flower), 1604c (S!).

J. dregeanus var. *conglomeratus* Buchen., l.c. 463. Type: Cape, Assegaaibos, *Ecklon & Zeyher* 10 (BOL!; W!).

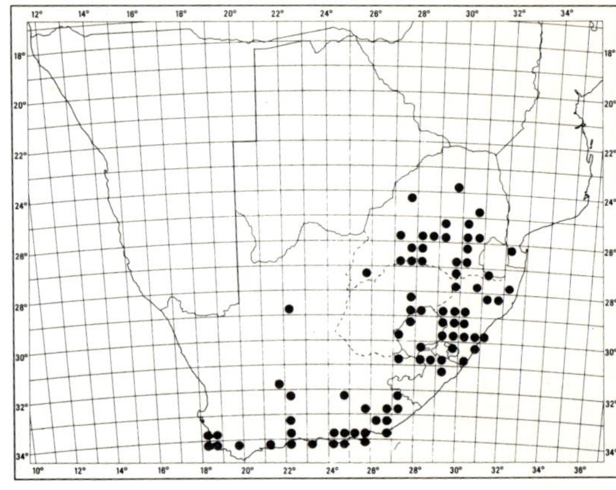

MAP 67.— **Juncus dregeanus**

J. sonderianus Buchen., l.c. 476 (1875) and in Pflanzenreich 4, 36: 252 (1906); Bak. in F.C. 7: 25 (1897); Adamson in J. Linn. Soc., Bot. 50: 25 (1935). Syntypes: Cape, Port Elizabeth on sand hills, *Drège* e (G!; K!; LD!; W!, PRE, photo.!); on dunes near Cape Recife, *Ecklon & Zeyher* 9, 780 (BOL!; W!).

J. dregeanus var. *sphaerocephalus* Adamson in J. Linn. Soc., Bot. 50: 21 (1935). Type: Cape, Cape Flats, *Adamson* 234 (BOL, lecto.!).

J. subcuneatus Adamson in J. Linn. Soc., Bot. 50: 23 (1935). Type not indicated.

J. subcuneatus var. *latifolius* Adamson, l.c. Type: Transvaal, Carolina, *Rogers* 21308 (J, iso.!).

J. subcuneatus var. *minor* Adamson, l.c. Type: Transvaal, Johannesburg, *Moss* 7986 (J, holo.!).

J. subglobosus Adamson in J. Linn. Soc., Bot. 50: 24 (1935). Type: Transvaal, Florida, *Moss* 7922 (J, holo.!).

Compact, tufted perennials, often turning brown when dry, c. 0,1–0,4 m tall. *Rhizome* small; roots thin, numerous. *Leaves* grass-like, narrowly linear-acuminate, with inrolled margins to filiform, erect, margins narrowly membranous below, initially with a closed basal sheath, not auriculate. *Inflorescence* consisting of crowded (rarely solitary) capitula which are sessile, or sometimes on short lateral branchlets, the young heads semiglobose, becoming globose and compact with age, usually exserted on long peduncles; lowest bract leaf-like but reduced; floral bracts small, membranous. *Tepals* usually dark, 3–4 mm long, outer aristate, inner acute with broad membranous margins, frayed with age. *Stamens* 6 (3–4), less than half as long as perianth; anthers small. *Ovary* trilocular, globose; style very short; stigmas red, curling up on top of ovary. *Capsule* globose, obtuse, tricostate, 3 mm long, firm, shiny brown with wide wavy placentas; seeds minute, 0,25 mm diam., globose, reticulate, ale-coloured with a small black apiculus.

Widely distributed in the eastern parts of Africa; common and widespread in Southern Africa from Transvaal through Natal and Lesotho to the south-western Cape. In wet localities, near the sea, to high altitudes. Flowering in summer. Recorded as an adventive in New Zealand. Map 67.

The usually compact, shortly bristly, dark capitula, the short style and small stamens are typical of the species. Flowers examined bore 6 stamens, but it is reported that only 3 were seen in some plants.

J. sonderianus Buchen., placed in synonymy under this species, may possibly be a hybrid between *J. dregeanus* and *J. lomatophyllus*. Several collections, mostly from around Port Elizabeth, possess leaves similar to those of *J. lomatophyllus* (no. 13) but have a compact globose inflorescence resembling that of *J. dregeanus*.

Vouchers: *Adamson* 257; *Lubke* 288a; *Schlechter* 3945; *Venter* 1070; *Werdermann & Oberdieck* 2131.

16. **Juncus cephalotes** *Thunb.*, Prodr. 1: 66 (1794) and Fl. Cap. 2: 337 (1823) pro parte; Buchen. in Abh. naturw. Ver. Bremen 4: 451, t. 7 (1875); Bak. in F.C. 7: 24 (1897); Adamson in J. Linn. Soc., Bot. 50: 32 (1935). Type: Cape, *Thunberg* 8708 (UPS, lecto.).

J. pictus Steud., Syn. Pl. Glum. 2: 305 (1855); Buchen., l.c. 458 (1875); Bak. in F.C. 7: 23 (1897); Adamson in J. Linn. Soc., Bot. 50: 35 (1935). Type: Cape, Namaqualand, Leliefontein, *Drège* 2472a (BOL!; G!).

J. altus Buchen., l.c. 457 (1875); Bak. in F.C. 7: 24 (1897). — var. *altus* (Buchen.) Adamson in J. Linn. Soc., Bot. 50: 34 (1935). Type: Cape, Kogmanskloof, *Zeyher* 15 (BOL!; G!).

J. cephalotes var. *ustulatus* Buchen., l.c., 451 (1875). Syntypes: Cape, Table Mountain, *Ecklon* 13 (BOL!); *Ecklon* 901 (LD!).

J. cephalotes var. *varius* Buchen., l.c. Type: Cape, Camps Bay, *Ecklon* s.n. (BOL!).

J. inaequalis Buchen., l.c. 455, t. 7 (1875); Bak. in F.C. 7: 24 (1897); Adamson in J. Linn. Soc., Bot. 50: 33 (1935). Syntypes: Cape Town, Camps Bay, *Ecklon* 24 (BOL!); *Ecklon* 12 (BOL!).

J. inaequalis var. *viridescens* Buchen., l.c. Type: Cape, Swellendam, *Zeyher* 4319 (BOL!; W!) distributed as *J. isolepoides* N. ab F. in sched. et in Linnaea 20: 244 (1847), pro parte, nom. nud.

J. parvulus E. Mey. ex Buchen., l.c. 447, t. 6 (1875); Bak. in F.C. 7: 22 (1897); Adamson in J. Linn. Soc., Bot. 50: 35 (1935). Type: Cape, Namaqualand, Modderfontein, *Drège* 2472b (S, holo.; BOL, iso.!).

J. polytrichos E. Mey. ex Buchen., l.c. 448, t. 6 (1875); Bak. in F.C. 7: 23 (1897); Adamson in J. Linn. Soc., Bot. 50: 35 (1935). Type: Cape, Leliefontein, *Drège* 2472aa (G!; K!).

J. schlechteri Buchen. in Bot. Jb. 459 (1898) and in Natürl. PflFam. 4, 36: 262 (1906). *J. rupestris* var. *schlech-*

teri (Buchen.) Adamson in J. Linn. Soc., Bot. 50: 38 (1935). Type: Cape, Bain's Kloof, *Schlechter* 9154 (BOL!; PRE!).

J. filifolius Adamson in J. Linn. Soc., Bot. 50: 36 (1935). Type: Cape, vlei on Cape Flats, *Adamson* 212 (BOL, holo.!).

J. inaequalis var. *squarrosus* Adamson in J. Linn. Soc., Bot. 50: 33 (1935). Type: Cape, Saron, *Schlechter* 10608 (PRE, iso.!).

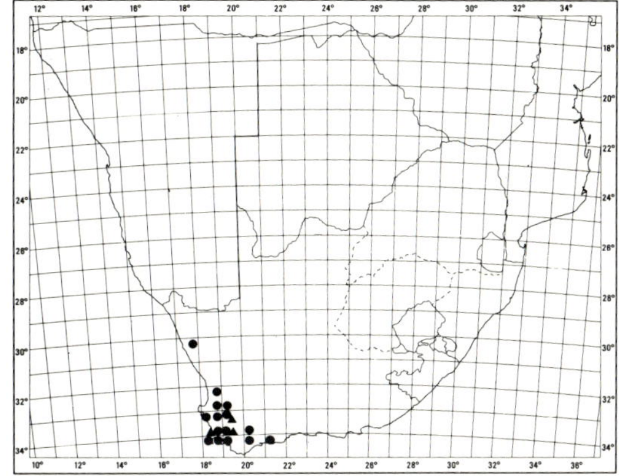

Map 68.— ● **Juncus cephalotes**
▲ **Juncus stenopetalus**

Annuals 0,1–0,15(–0,3) m tall, tufted; epidermis verruculose. *Leaves* linear-acuminate, 40–80 × 1–3 mm, flat, soft, with a reddish tinge below. *Inflorescence* on an exserted ribbed peduncle; umbel compound, 4–6 erect side branchlets 10–30 mm long, arising from below solitary terminal semi-globose capitulum, occasionally doubly compound; in paedogenic seedlings with a solitary few-flowered capitulum. *Flowers* with tepals c. 3 mm long, usually streaked with shiny blackish brown or fawn, and broad central band colourless and scabrid, fading with age, outer tepals narrow, aristate, inner ones broader, with a wide membranous margin usually overtopping outer ones. *Stamens* 6; anthers long; filaments short. *Ovary* ovoid; style 1 mm long; stigmas red, much exserted, 3–5 mm long. *Capsule* oblong-globose, 3 mm long, obtuse, apiculate, thin-walled, pale; seeds narrowly ellipsoid, asymmetrical, 0,5 mm long, apiculate at apex and base, brown; testa white, membranous, hygroscopic. Fig. 25: 3.

Distributed in the south-western Cape from Namaqualand to the Swellendam District. Flowering in spring. Map 68.

FIG. 25.—1, **Juncus capensis**: 1a, habit, × 0,3; 1b, gynoecium, × 10 (*Acocks* 23115). 2, **J. dregeanus**: 2a, habit, × 0,5; 2b, young fruit, × 10 (*Venter* 1071). 3, **J. cephalotes**, habit (small specimen), × 1 (*Adamson* 333). 4, **J. lomatophyllus**: 4a, basal part of plant, × 0,3; 4b, inflorescence, × 1; 4c, stamens and tepal, × 10 (*Goldblatt* 6429).

Vouchers: *Adamson* 207; *Esterhuysen* 6123; *Liebenberg* 6541; *Parker* 3730.

Characteristic of *J. cephalotes* is the minutely verrucose epidermis of the stem. The vascular bundles with their large thickened cells form prominent colourless ridges. The broad midrib of the tepals is also distinctly verruculose, often pale. The drawings by Buchenau (1875: t. 7) do not show the full length of the stigmas.

17. Juncus stenopetalus *Adamson* in Jl S. Afr. Bot. 8: 273 (1942), nom. nov. for *J. sprengelii* Nees ex Buchen., non Willd. (1787). *J. sprengelii* Nees in Linnaea 20: 244 (1847), nom. nud.; Buchen. in Abh. naturw. Ver. Bremen 4: 449 (1875), including vars *robustior* Buchen. and *gracilior* Buchen.; Bak. in F.C. 7: 24 (1897); Adamson in J. Linn. Soc., Bot. 50: 34 (1935). Type: Cape, Tulbagh Waterfall, *Ecklon & Zeyher* 11 (BOL!; LD, iso.!).

Small annuals c. 0,14 m tall. *Leaves* 1 to few per shoot, narrowly linear-acuminate, c. 80 mm long, flat, epidermis reticulate, margin membranous below. *Capitulum* 1, rarely 2–3, apical, globose, 6–12-flowered; peduncle minutely scabrid. *Flowers* fusiform, on short pedicels. *Tepals* narrowly acuminate, c. 6 mm long, costate, minutely scabrid. *Stamens* 6, about 2 mm long; filaments very short. *Ovary* ovoid-acuminate with a long style and stigmas. *Capsule* ovoid, attenuated into a long beak; seeds oblong-globose, c. 0,4 mm long, reticulate, apex dark.

Rare, collected in the Cape near Camps Bay, Tulbagh Waterfall and Bain's Kloof. Map 68.

Vouchers: *Compton* 18645 (NBG); *Ecklon & Zeyher* 11; *Schlechter* 9129.

Closely related to *J. cephalotes* Thunb. (above).

18. Juncus rupestris *Kunth*, Enum. Pl. 3: 344 (1841); Buchen. In Abh. naturw. Ver. Bremen 4: 441 (1875); Bak. in F.C. 7: 21 (1897); Adamson in J. Linn. Soc., Bot. 50: 37 (1935). Type: Cape, Kamiesberge, Eselsfontein, *Drège* 2471a (G!; K!; LD!; BOL, iso.!, PRE, photo.!).

Small delicate annuals 60–160 mm tall, forming small tufts; stems verruculose. *Leaves* many, filiform, 20–40 × 0,5 mm, reddish below, apiculate. *Inflorescence* umbellate, side branches 1–3 (–5), filiform, up to 15 mm long, formed below oldest flowers, occasionally branching once again or, in young plants, inflorescence simple; capitula 1–4-flowered. *Flowers* ovoid, 3 mm long; outer tepals aristate, inner broader, obtuse, with a broad

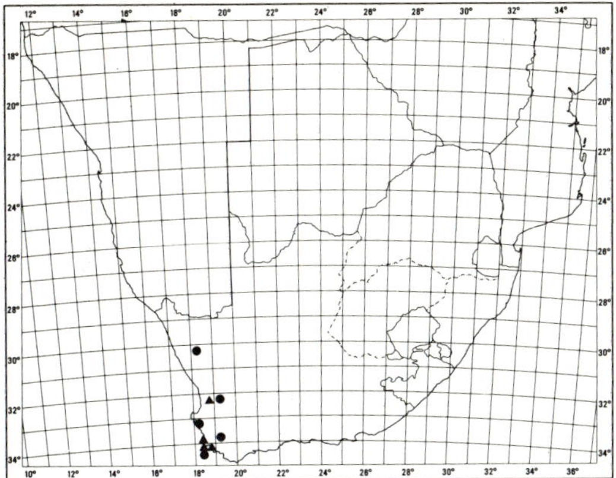

MAP 69.— ● **Juncus rupestris**
       ▲ **Juncus scabriusculus**

membranous margin, about as long as outer. *Stamens* 6, short; anthers about equal to filaments in length. *Ovary* oblong-globose, 3-locular; style very short; stigmas 3, red, very short. *Capsule* oblong-globose, obtuse; seeds globose, 0,25 mm diam., golden yellow with a minute apiculus, obtuse to indented at base, indistinctly reticulate.

Endemic to the south-western Cape; not often collected, probably because of its small size. Flowering in spring. Map 69.

Vouchers: *Acocks* 17421, 23414; *Esterhuysen* 12083; *Schlechter* 9224; *Wolley Dod* 3400.

19. Juncus scabriusculus *Kunth*, Enum. Pl. 3: 354 (1841); Buchen. in Abh. naturw. Ver. Bremen 4: 444, t. 6 (1875); Bak. in F.C. 7: 22 (1897); Adamson in J. Linn. Soc., Bot. 50: 36 (1935). Type: Cape, Piketberg Range, near Groene Vallei, *Drège* 8795 (G!, K, isosyn.!; PRE, photo.!).

J. subglandulosus Steud., Syn. Pl. Glum. 2: 303 (1855); Buchen., l.c. 459, t. 6 (1875); Adamson in Jl S. Afr. Bot. 3: 167 (1937).

J. scabriusculus var. *subglandulosus* (Steud.) Buchen. in Bot. Jb. 12: 458 (1890); Bak. in F.C. 7: 22 (1897). Type: Cape, Piketberg Range, near Groene Vallei, *Drège* 8795 pro parte. According to Baker, l.c. Buchenau's locality of the *Drège* type, viz. Witte Bergen near Aliwal North, appears to be an error.

Small annuals 60–120 (–300) mm tall. *Leaves* basal, narrowly linear, 40–80 mm long, c. 2 mm broad, flat or folded. *Inflorescences* on erect peduncles just exserted from leaves. *Capitulum* apical, solitary or rarely with 1–2 smaller ones above, cup-shaped, compact, basal bracts short, resembling tepals, glumaceous; outer and inner tepals subequal, 6 mm long, aristate. *Stamens* 6, less than half as long as tepals; filaments equalling anthers. *Ovary* cylindrical-triangular, trilocular; style very short with 3 stigmas c. 3 mm long. *Capsule* cylindrical-triangular, shorter than tepals, obtuse, apiculate; placentas 3, persisting as a central column after dehiscence; seeds oblong, 0,75 mm long, obtuse, reticulate, brownish orange. Plate 3: 1.

South-western Cape: recorded from the Peninsula, Paarl and Bredasdorp; in damp localities. Flowering October. Map 69. Plate 3: 1.

Vouchers: *Acocks* 22771; *Adamson* 331; *H. Bolus* 4812.

20. **Juncus capitatus** *Weigel*, Obs. Bot. 28, t. 2, fig. 5 (1772); Buchen. in Pflanzenreich 4, 36 (Heft 25): 256, fig. 24 (1906); Adamson in J. Linn. Soc., Bot. 50: 32 (1935). Cutler in Anat. Monocot. 4: 38, 39, 72, 74 (1969). Type: Germany, *Wilke*.

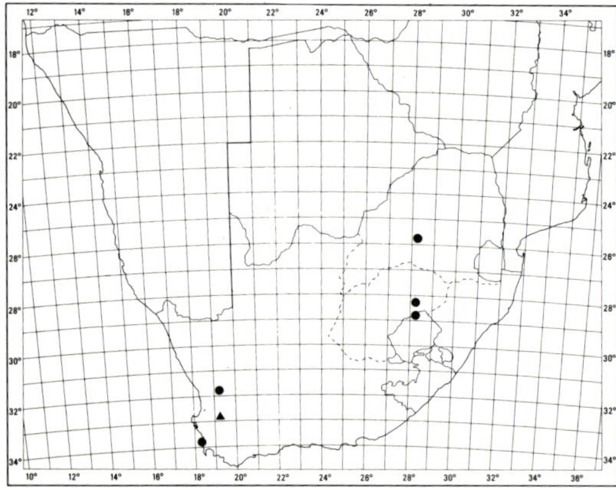

Map 70.— ● **Juncus capitatus**
▲ **Juncus obliquus**

Very small tufted annuals (20–) 50–80 mm tall. *Leaves* several per shoot, narrowly linear, about half as long as inflorescences, forming a broad, membranous, folded sheath basally. *Capitula* single, apical, pseudolateral, 2–8-flowered, lowest bract leaf-like forming continuation of stem, exserted above. *Flowers* chasmogamous or cleistogamous. *Tepals* 3–4 mm long, ovate, transparent, pale, often pinkish, outer aristate, inner shorter, apiculate. *Stamens* 3, opposite outer tepals; filaments longer than anthers. *Ovary* ovoid; style long, with erect stigmas or short, with stigmas curled up on top of ovary. *Cap-*

sule trilocular, ovoid, mucronate, included in perianth, shiny, brown; seeds 0,3 mm long, obovate, reticulate, golden brown. Plate 3: 2.

A widespread, cosmopolitan annual. Rare in Southern Africa where it has been recorded from Transvaal, Orange Free State, Lesotho and Cape. Map 70.

Vouchers: *Acocks* 17422; *Dieterlen* 806; *Esterhuysen* 32932; *Mauve & Venter* 5068; *Moss & Otley* 11949.

Buchenau (l.c., fig. 24; 1906) believes the cleistogamous, short-styled flowers to be most probably self-pollinated.

21. **Juncus obliquus** *Adamson* in Jl S. Afr. Bot. 3: 165 (1937). Type: Cape, Ceres Div., Cold Bokkeveld Mts, Ebenezer, *Adamson* 1139 (BOL, holo.).

Annuals c. 50 mm tall, delicate, tufted. *Leaves* c. 4 per plant, setaceous, 10–15 mm long, erect, broadened below, sheathing peduncles. *Capitula* solitary, pseudolateral, rarely a second one produced above, 1–4-flowered; bracts tepaloid, lowest forming continuation of stem but not overtopping flowers. *Tepals* ovate-concave, 2,5 mm long with dark brown margins, outer shorter, pointed. *Stamens* 3, about half as long as tepals; anthers somewhat shorter than filaments. *Ovary* ovoid; style short; stigmas fairly long, spreading. *Capsule* oblong, obtuse, apiculate, golden brown; seeds ovoid, obtuse, 0,3 mm long, golden brown, reticulate.

Known only from the Cold Bokkeveld Mts in the Ceres District. In flower and fruit September–October. Map 70.

Vouchers: *Adamson* 2998; *Levyns* 5791.

This diminutive species is perhaps closest to *J. rupestris* (no. 18), but differs in possessing only 3 stamens, a short style and usually a single head. The three known collections are almost identical.

Species insufficiently known

Juncus diaphanus Buchen. in Abh. naturw. Ver. Brem. 4: 442 (1875); Adamson in J. Linn. Soc., Bot. 50: 37 (1935). Type: Cape, Albany, *Bolus* 188 (in Sonder herbarium). This specimen was not found in the Buchenau herbarium at Vienna. *Bolus* 188 in BOL from the Sondags River at Graaff-Reinet is *J. exsertus*; cf. Buchenau in Abh. naturw. Ver. Bremen 4: 435 (1875). Buchenau, in a letter attached to a sheet of *J. scabriusculus* in BOL, suggested that *J. diaphanus* could perhaps belong to *J. scabriusculus*, but this latter species only occurs in the southwestern Cape. The compact heads on short branchlets, the small anthers, short style and obtuse capsule, as well as the broad leaves, all suggest that it could be *J. dregeanus* Kunth, which is common in the eastern Cape.

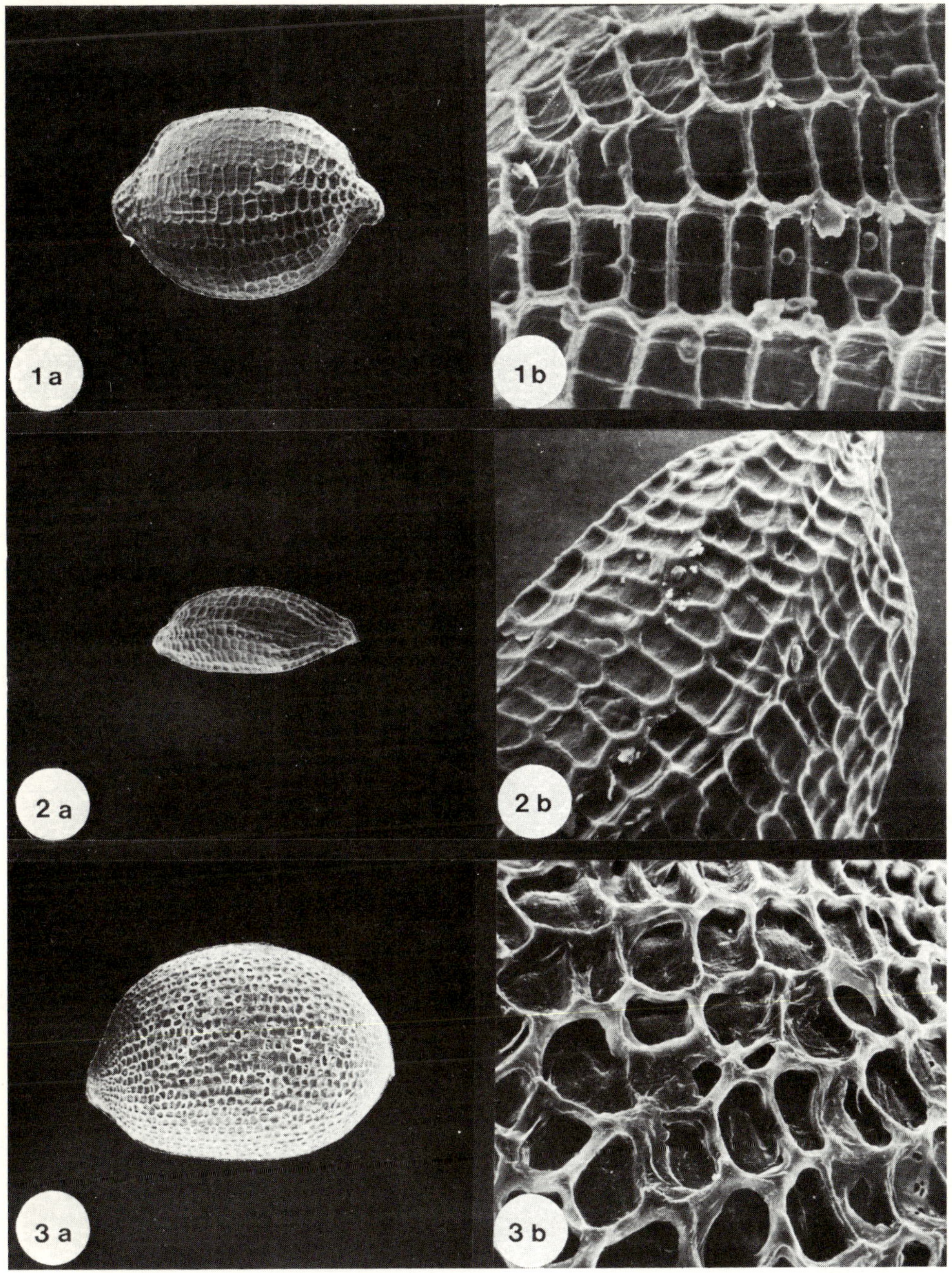

Plate 3.—Seeds: 1, **Juncus scabriusculus**: 1a, × 120; 1b, × 1200 (*Adamson* 206). 2, **J. capitatus:** 2a, × 120; 2b, × 600 (*Esterhuysen* 32932). 3, **Luzula africana**: 3a, × 240; 3b, × 600 (*Arnold* 851).

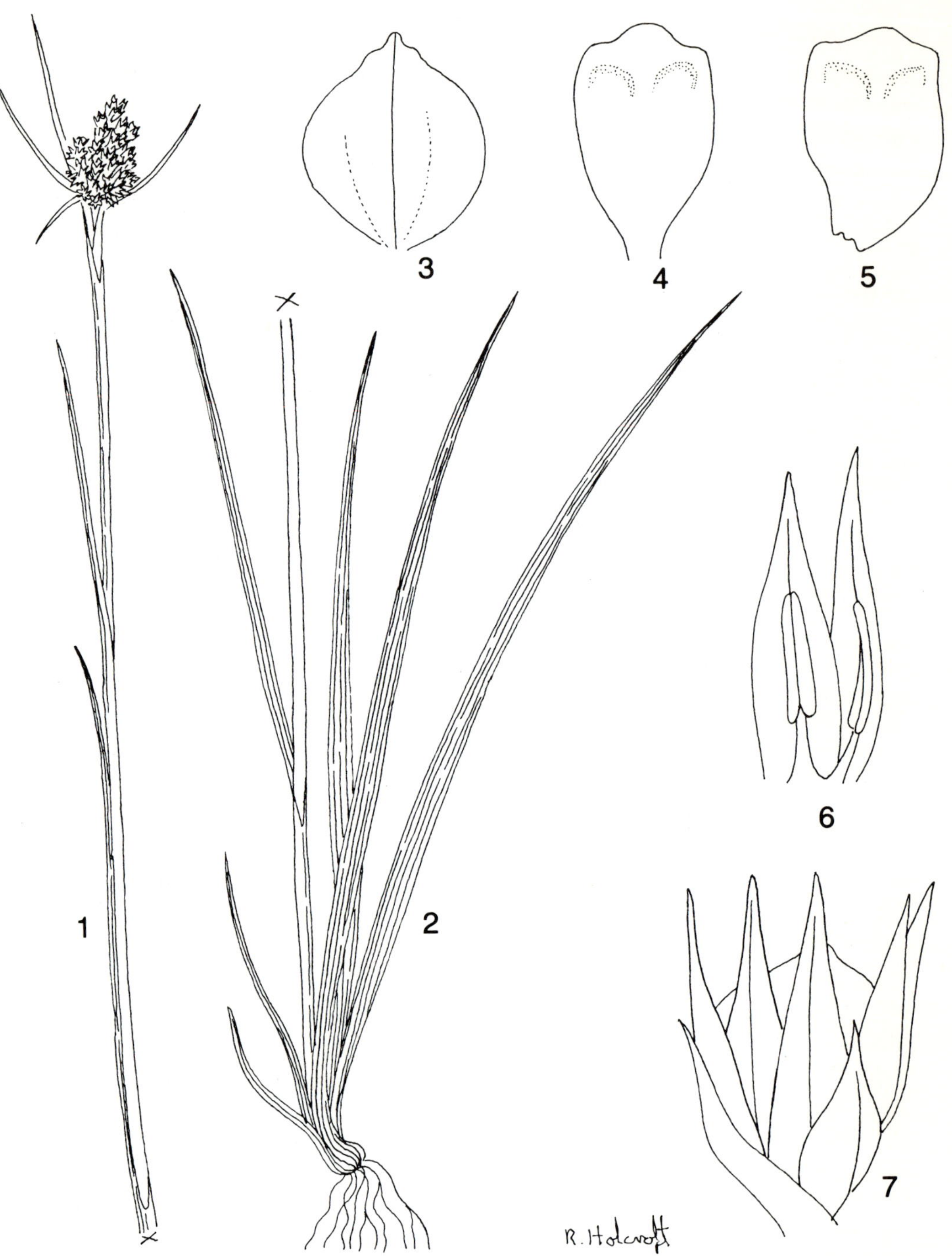

FIG. 26.—**Luzula africana**: 1 & 2, whole plant, × 1; 3, capsule, × 10; 4, seed, ventral view, × 20; 5, seed, lateral view, × 20; 6, tepals and stamens, × 10; 7, flower with ripe capsule and bracts, × 10; (1 & 2, after *O. West* 1683; 3–7, after Buchenau).

937   **3. LUZULA**

Luzula *DC.* in Lam. & DC., Fl. Franc. 1: 198 (1805) & 3: 158 (1805), nom. cons.: Buchen. in Pflanzenreich 4, 36 (Heft 25): 42 (1906); Bak. in F.C. 7: 27 (1897); M. Friedrich et al. in F.S.W.A. 156: 3 (1967); R. A. Dyer, Gen. 2: 915 (1976). Type species: *L. campestris* (L.) DC.

Tufted, grass-like annuals or usually perennial herbs. *Leaves* flat, linear, with closed basal sheaths, inconspicuously silky-hairy. *Inflorescence* of dense, terminal spikes, heads or panicles, rarely few-flowered. *Scapes* long, furnished with reduced leaves or bracts, upper ones subtending inflorescence. *Flowers* bisexual, small; perianth of 6 free segments, glumaceous. *Stamens* 3 (6). *Ovary* unilocular, with 3 nearly basal ovules. *Capsule* loculicidal; seeds ovoid to globose, with or without an apical or basal appendage.

Species about 80, mostly north temperate regions, rare in Southern Hemisphere; one species in Southern Africa.

The name *Luzula* alludes to the shiny leaves and flowers of some species.

Luzula africana *Drège ex Steud.*, Syn. Pl. Glum. 2: 294 (1855); Buchen. in Abh. naturw. Ver. Bremen 4: 414 (1875); Bak. in F.C. 7: 27 (1897); M. Friedrich et al. in F.S.W.A. 156: 2 (1967). Type: E. Cape, Katberg, *Drège* 3963 (G!; LD, iso.!).

Small, tufted, grass-like herbs, 0,3–0,4 m high. *Rhizome* compact, small, fibrous; roots numerous, thin, branched. *Leaves* basal, several, erect, narrowly linear, 0,1–0,2 m long, soft, with scattered, long, fine, white hairs along raised margin, apex callose, obtuse, basal sheath cylindrical. *Stem* erect, thin, with few reduced leaves. *Inflorescence* an apical, congested head, subtended by 1–few leaf-like bracts. *Flowers* small, brownish, each subtended by a small membranous, ciliate bract. *Tepals* lanceolate-acuminate c. 2,5 mm long, inner ones slightly shorter, membranous. *Stamens* 6, c. 2 mm long; filaments longer than anthers. *Ovary* ovoid-acute, unilocular; style short, with 3 longer stigmatic branches; ovules 3, basal, erect. *Capsule* globose, apiculate; seeds globose, c. 1 mm long, dark, minutely apiculate, with a thick, basal caruncula,

epidermis thick, membranous. Fig. 26. Plate 3: 3.

Found at high altitudes in moist grassveld and marshes on the eastern Transvaal Highveld, the Drakensberg in Natal, eastern Orange Free State, Lesotho and northeastern Cape. Once recorded from South West Africa/Namibia. Map 71. Plate 3:3.

Vouchers: *Arnold* 851; *Dieterlen* 734; *Dyer* 777; *Killick & Marais* 2077; *Kinges* 3427; *Moll* 1202.

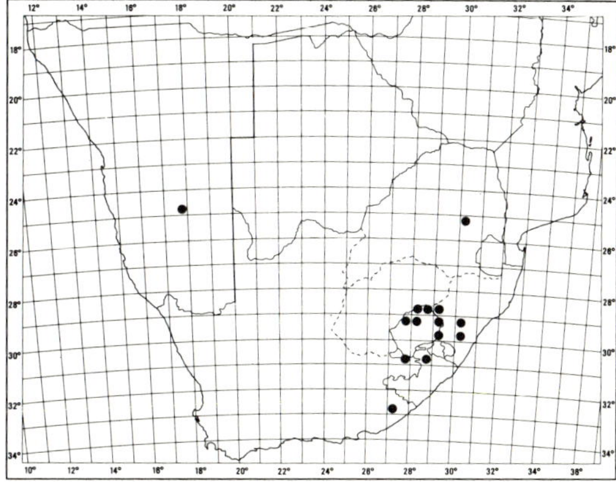

MAP 71.— **Luzula africana**

INDEX

---

* An asterisk signifies exotic genera and species which are not naturalized; synonyms are in italics.